OCEAN
ART

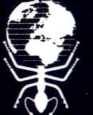

OCEAN
ART

FROM THE SHORE
TO THE DEEP

HELEN SCALES

For Riam and Eric

Published by
Reaktion Books Ltd
Unit 32, Waterside
44–48 Wharf Road
London N1 7UX, UK

www.reaktionbooks.co.uk

First published 2025
Copyright © Helen Scales 2025

EU GPSR Authorised Representative
Logos Europe, 9 rue Nicolas Poussin, 17000, La Rochelle, France
email: contact@logoseurope.eu

Printed and bound in India
by Replika Press Pvt. Ltd

A catalogue record for this book is available from the British Library

ISBN 978 1 83639 098 5

CONTENTS

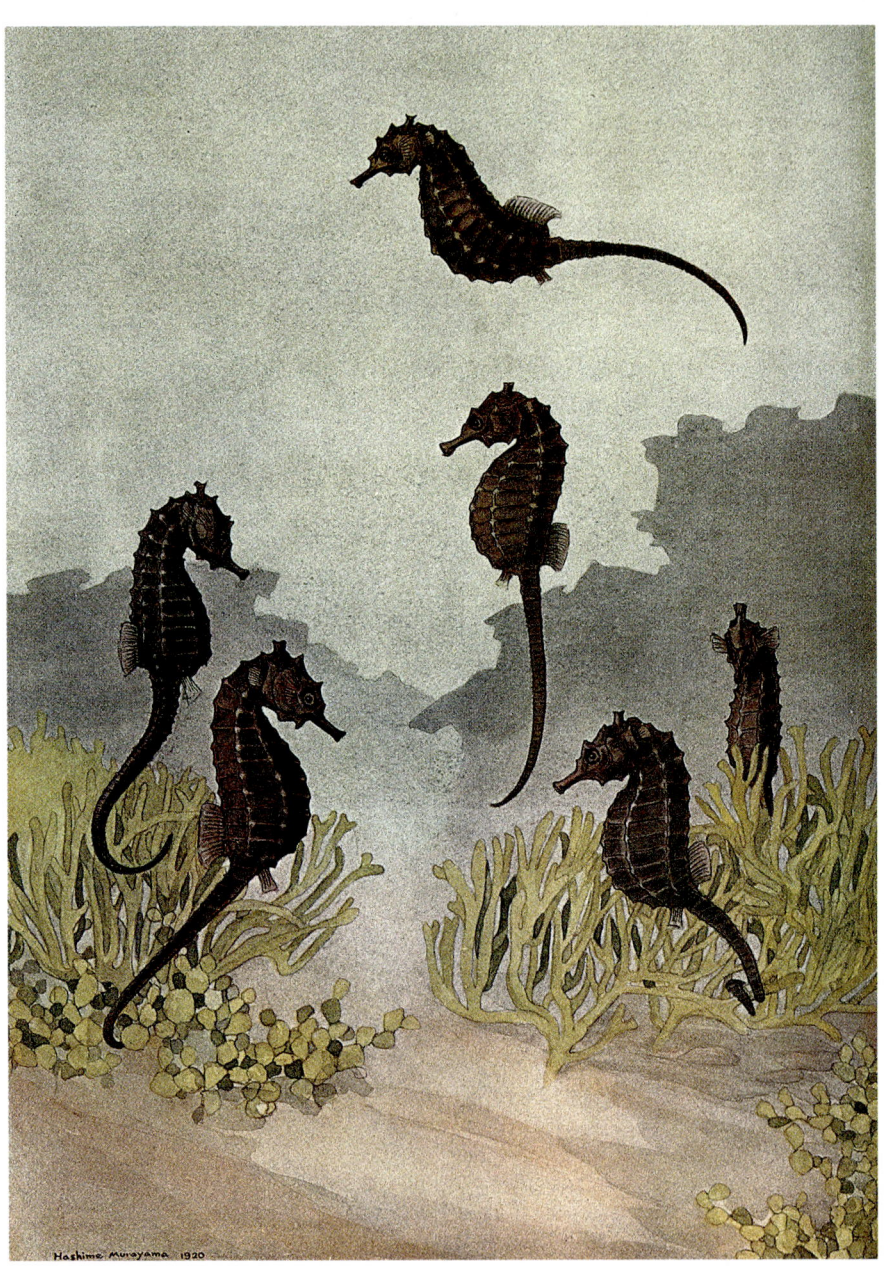

Hashime Murayama 1920

INTRODUCTION

Earth is a blue planet. Saltwater covers seven-tenths of the surface and swirls around the continents, giving the globe its sapphire complexion. Beneath the waterline, the seas go down and down, making this the single biggest portion of the planetary biosphere. Of all the space available for life to exist, more than 95 per cent is ocean.

All this water makes Earth habitable. Without it, life as we know it would be impossible and would likely never have got going in the first place. This is where, billions of years ago, the first living cells assembled, then through the aeons gradually gathered together, diversified, proliferated and grew into complex, multi-cellular beings. Everything around us on land originally came from the ocean; the simple plants that painted the coasts with an emerald fringe then moved inland; the ancestors of insects and worms that scuttled and crept across the tideline and kept on going. Our own backboned ocean-dwelling ancestors were drifters and swimmers, who in time evolved legs, first for paddling in the shallows, then for bearing their weight and lifting their bodies into the air. Long before there were humans, some vertebrates went back to the ocean, taking with them gulps of air and evolving their legs once again into swimming flippers and flukes.

Now here we all are, living among the ocean's distant progeny. It is easy to forget about the realm our forebears left behind and all the other life forms that dwell there now. Ocean life, for most people, most of the time, lies out of sight and out of mind. That blue blanket of the surface sea covers over and hides all that lies below. Of course, people go there to explore and study this crucial part of the planet – scuba divers, snorkellers, free divers and scientists like myself who feel the urge to be in the ocean, to learn about it and understand how the vast ecosystem works. For

'The seahorse (Hippocampus)', drawn from life by
Hashime Murayama, 1920.

humans, entering the ocean can be a great physical and mental challenge: to withstand the power of the waves, tides and currents; to keep warm in the heat-sapping water; to bear the hydrostatic pressure pressing down. For some, the hardest part is overcoming fears of the lurking, unseen beasts, and this alone is enough to keep plenty of people away. Explorers of many kinds follow their curiosity into the ocean and many have brought back observations, ideas and objects to show others what's down there and what life in the ocean is really like. The ocean also leaves abundant clues as to its living contents scattered along the tideline and even far inland from times when there was more ocean and less dry land. Depictions of sea life, in works of art, science and craft, as well as artefacts that come from the ocean, together make up a material culture that embraces, celebrates and seeks to comprehend ocean life.

I fell in love with the ocean a long time ago. This affinity began during childhood holidays at the seaside, sometimes on the Atlantic coast of France and most often on the wild beaches of Cornwall. And it was in Cornwall, in my late teens, that I first ventured way below the waterline and adopted a fish's perspective on the world. These were my training scuba dives as I learned how to cope with bulky diving equipment and very cold water, and I found within myself the ability to suppress panic when the visibility underwater shrank to less than an arm's length in the murky, plankton-filled seas. All of this was worthwhile for the extraordinary living wonders that it allowed me to spend time with – the cuttlefish and their ever-changing skin colours, the electric blue cuckoo wrasse, pencil-thin pipefish and elegant catsharks napping on the seabed.

My response to encountering these animals was to find out more about their lives and to follow them deeper into their world. And so, diving licence in hand, I went on to train to be a marine biologist. My university tutors showed me how to see the world in a certain way, to ask questions, gather data and observations and to build answers. Then off I set to explore more of the ocean and conduct my own studies, focusing on the changes that humans are imposing on sea life and seeking ways to undo the troubles of over-fishing and the damage wreaked on important, fragile habitats.

All the while, I have been constantly spellbound and inspired by the beauty of life in the ocean – the colours and patterns, shapes and movements, all so very different from what I see around me on land. Early on in my schooling, I was asked to make a choice between following an artistic life and following a scientific one – the two could not easily be combined. I loved art classes and making things, but left those interests behind when I picked my scientific route into the ocean. Maybe one day I'll go back and seek out some formal artistic tuition. For now, I occasionally dig out my art supplies, remind my hands what to do and make pictures that are just for those closest to me, those who know I find it hard to paint or draw anything else besides sea creatures. At the same time, I'm fascinated by others who have been captivated by

'The creatures of the sea are asked by the Ocean King to take a message to the Brahman', from a Tuti-Nama *(Tales of a Parrot), Mughal India, c. 1560, ink and gold on paper.*

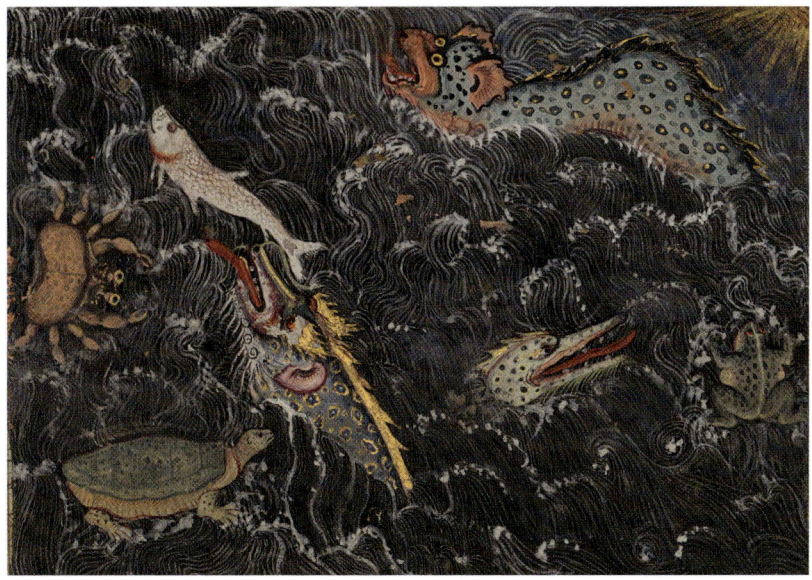

life in the ocean and who have found their own ways to portray these other-worldly life forms.

Bringing the ocean's hidden species into view as works of art makes them more relevant to people's lives, and harder to overlook and ignore; it shows that even though humans don't actually live in the ocean and most people, except for mariners, oceanographers and fishers, spend little or no time sailing across the ocean surface, this is still a place full of living marvels that are connected to life everywhere on Earth.

This book explores how people have sought out sea creatures – both the real living things and the imagined, mysterious beings that hold great power – and created images and objects to show others so they too can know and comprehend them. Think of the following chapters as a tour through museums and art galleries with a marine biologist as your guide, picking out their favourite images and artworks of sea life. I will be comparing these works to what I know of the ocean and reflecting on how they shift my own view of the underwater world. I want to show parts of the world that I adore – not just from my scientific viewpoint but through the eyes of artists and craftspeople who have their own reasons for looking into the ocean and seeing what they can find.

Life in the ocean has an ephemeral beauty that is very quickly lost when those living things are taken from their habitats. Colours fade. Delicate bodies collapse into shapeless goo. Often, all that is left is pallid ghosts of the bodies that once rippled and pulsed through the water, or raced at high speed with bodies like bullets of polished steel, or glittered with their own inbuilt lights, or sprouted from the seabed, growing into so many intricate aquatic forests and meadows. In my studies, I've spent hours watching fish on coral reefs, admiring their vivid colours. Then, when fishers hauled them from the water, I've seen the fishes' radiance fade before my eyes, their bright scales dimming within minutes.

Collectors and scientists do their best to gather specimens of ocean-going species and preserve their bodies in as faithful a form as they can, both for academic studies and simply to know what

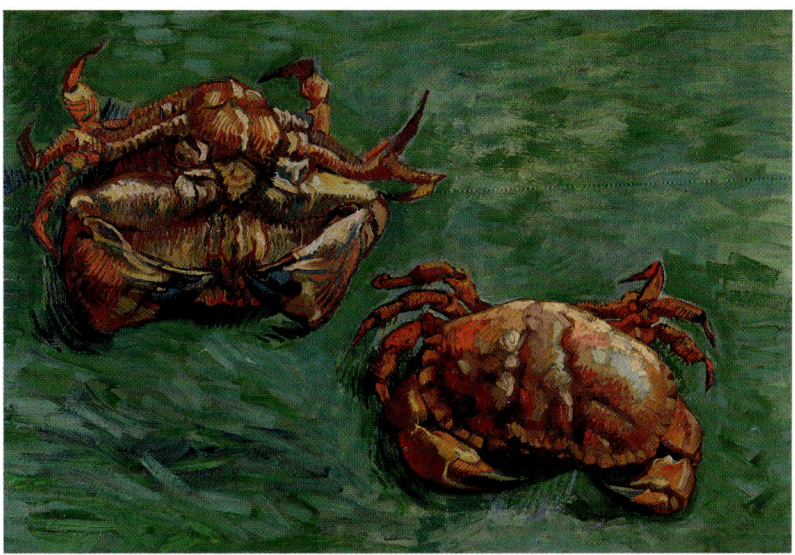

Vincent van Gogh, Two Crabs, *1889, oil on canvas.*

these creatures look like. Most often, sea creatures are removed from their original aquatic world and transferred to another captive ocean of preserving fluids (often in a two-step process, dunking first in formalin, which stops their tissues from breaking down, then immersing in ethanol, which keeps decomposing microbes at bay). Then once again the animals are afloat, but they are never quite the same as the wild, living bodies. The colours of the most vibrant coral reef species will quickly seep out, sometimes colouring their surrounding liquid (imagine if this happened out in the ocean, and the colours of creatures washed out). Soft, boneless invertebrates sink to the bottom of their eternal, confined chambers.

Some historic fish collections in museums are remarkably colourful thanks to the careful work of past curators who painstakingly painted the preserved animals, scale by scale. How they did this in some cases remains a mystery. At the University Museum of Bergen in Norway, today's curators are at a loss to know how their preserved fish remain so bright after decades – their predecessors didn't pass on details of the painting techniques they used.

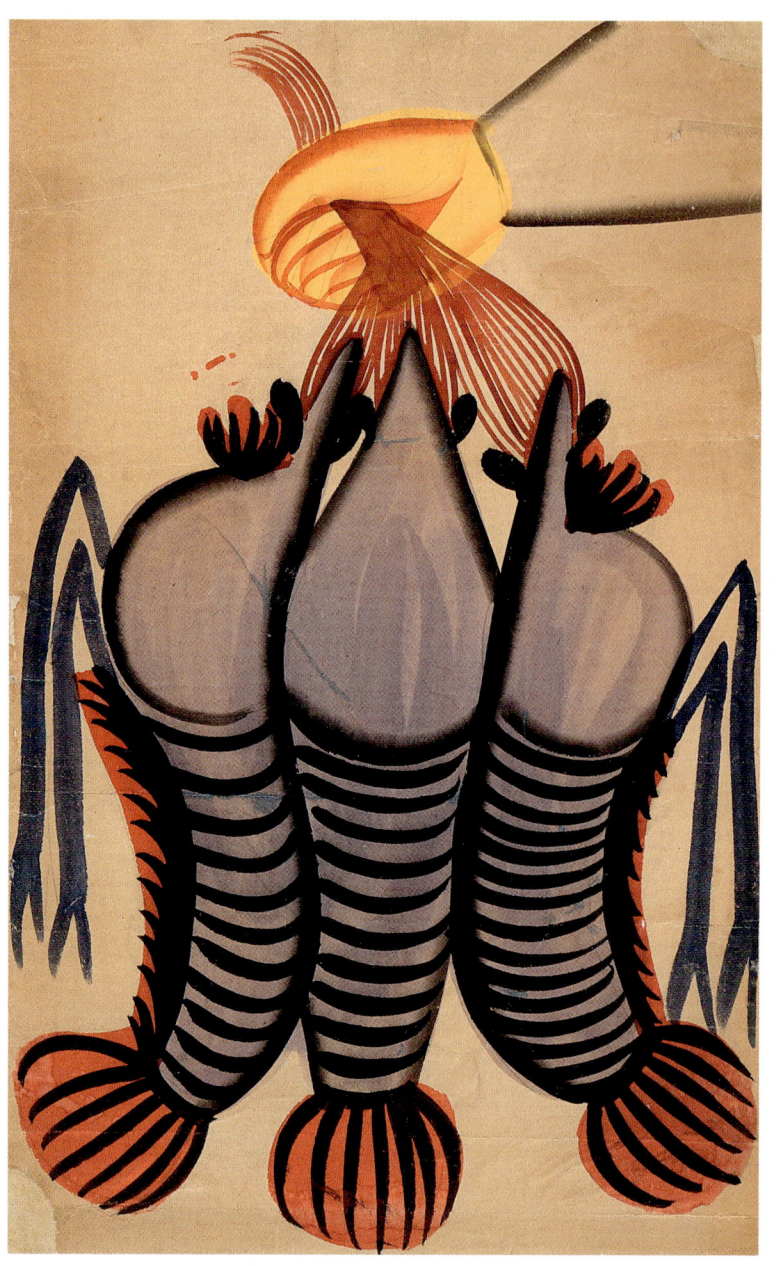

Fish taxidermy is popular with anglers who have their prize catches preserved and painted to fix on the wall for all to see. It is rarer to see museum collections with stuffed fish on display. In a visit to a small museum at a marine research station in western France I came across an impressive collection created by one of the only living ichthyotaxidermists, Bernard Bourlès. Walking through the fish gallery, I felt like I was diving in a strange mixed shoal of thresher sharks, basking sharks, Greenland sharks, sting-rays and sawfish, all animals that I would never see together in the ocean but here in the museum they were together frozen in a moment, as if ready to swim away.

Long before humans utilized scientific advances to preserve the bodies of fish, some very ancient sea creatures were preserved without human help, their bodies enveloped in sediments and over aeons turned to stone. They too, like the museum specimens, are not what they were, and require experts to interpret the identity

Opposite: *A hand holding prawns, India, West Bengal, second half of the 19th century, watercolour.* Below: *Kawahara Keiga,* Starry Handfish (Halieutaea stellata*), 1823–9, pencil and watercolour on paper.*

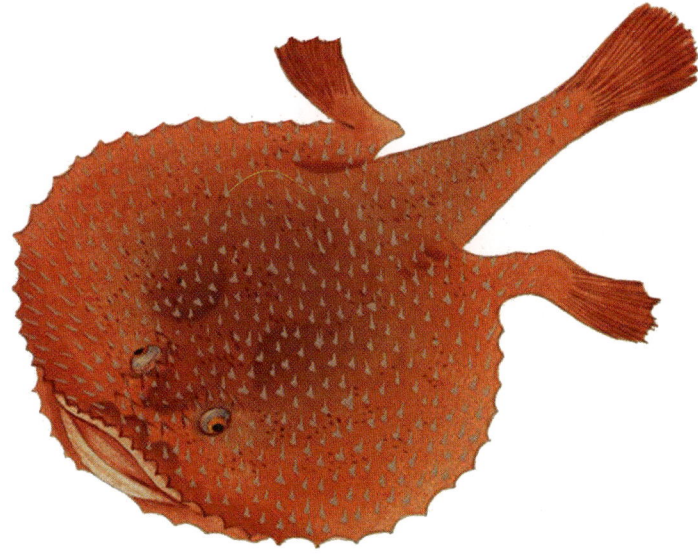

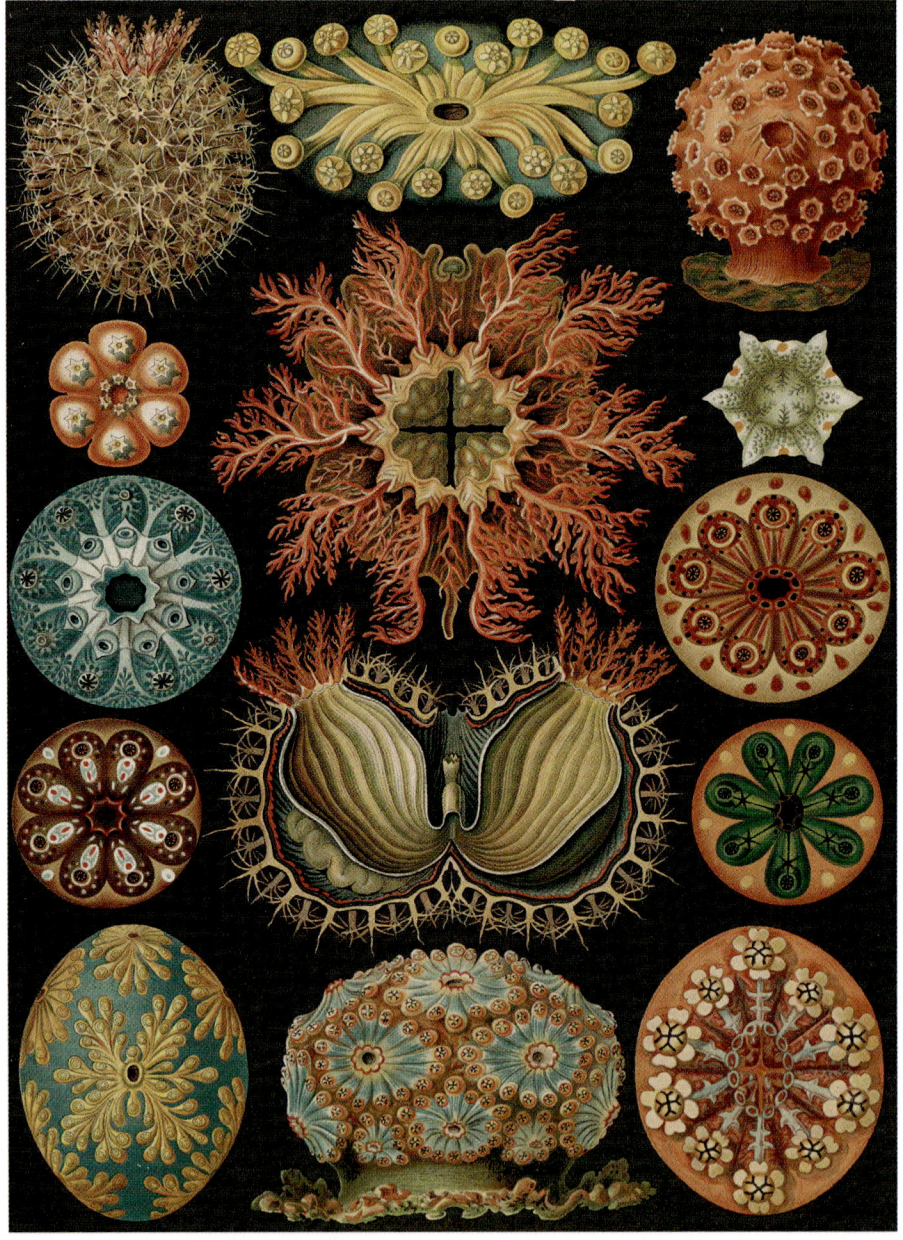

of the original animals. These enigmatic beasts have inspired beliefs and stories, and people have held onto them as powerful objects.

Beyond the exceptional skills of human and geological preservation, and until people built aquariums and then cameras – and worked out how to make them waterproof and pressure-proof – the only way to truly know what the seas' inhabitants look like was to go there and visit them in their world, and for most people throughout most of human history that hasn't been a possibility. Some artists in the past have stationed themselves on coastlines to be there when fishers brought back their catches so that they could see animals not long dead with their colours and shapes still largely intact. Some went to sea with fishers and scientists, and saw animals even more freshly captured. But for many, they would never lay eyes on a living or even recently deceased animal and relied instead on preserved specimens – the shrivelled, lustreless bodies that were passed onto them. And some artists never went near anything that came from the ocean and instead reproduced images from drawings and paintings made by others. Individual, distinctive sea creatures have appeared time and again in illustrations made by artists who copied existing images.

This book will explore many of the different ways that artists and craftspeople have found to reproduce and represent the ocean's living wonders and hold onto that fleeting beauty. These works are their own acts of preservation and transmission, passing on sea creatures through the ages.

A few things to note before we begin. While I have a great fondness for watching the waves from a sailboat or gazing down onto the sea from a high clifftop, the works I've picked out for this book are not those that aim exclusively to portray the physical nature of the ocean. For the most part, these are not the seagoing or maritime equivalents of landscape paintings. We won't be staring into the wild waves and contemplating great sea storms and

Sea squirts (ascidians), from Ernst Haeckel, Kunstformen der Natur *(1904).*

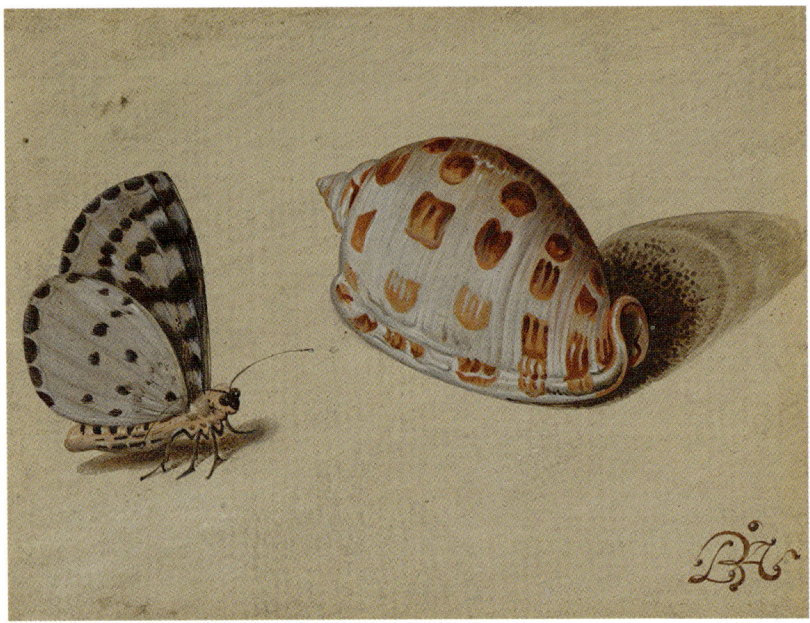

currents, but focusing on the living inhabitants of the ocean, or, as we'll see, ones that lived there long ago.

My aim here is to find works that celebrate the diversity of life in the sea. The ocean is home to myriad species, most of which have not yet been seen by human eyes, let alone collected, photographed, drawn and given an official name. Artists and scientists working together, or sometimes a single person embodying both professions, have played a critical part in describing and recording the ocean's immense biodiversity. Many of the works that I've selected here speak to this desire to know all the things that live in the seas – a task that is unlikely ever to be completed because of its sheer scale. At present, there are roughly 240,000 officially recognized entries in the World Register of Marine Species, a global clearing house for ocean taxonomists to aggregate their findings. This is as close as it is possible to get to a comprehensive list of known ocean species. By various calculations, experts generally agree that there are likely to be at least 2 million species alive today in the ocean. At the current pace of discovery, it will take hundreds of years to find them all and give each one a scientific name and description. Even with emerging technologies such as those that allow scientists to catch fragments of DNA from the water and determine which species swam nearby within the past few days, it's my guess that people will never produce a complete catalogue of ocean life, and there will always be the possibility that we've missed something in this enormous space.

Today, drawings are still an important part of the work of taxonomists. I spoke recently to an ichthyologist who had a pre-served specimen of a female deep-sea anglerfish, a black-bodied fish with a huge mouth full of teeth and a prong on her forehead that would have glowed at the tip while she was still alive. He sus-pected that she was something different from other, known species, and in order to find out he first needed to combine his

Opposite: *Circle of Balthasar van der Ast,* An Arrowhead Blue Butterfly and a Scotch Bonnet Sea Shell, *17th century, oil on paper.* Below: *Peter Paillou the Younger (active 1757–1831),* Lumpsucker, *n.d., watercolour.*

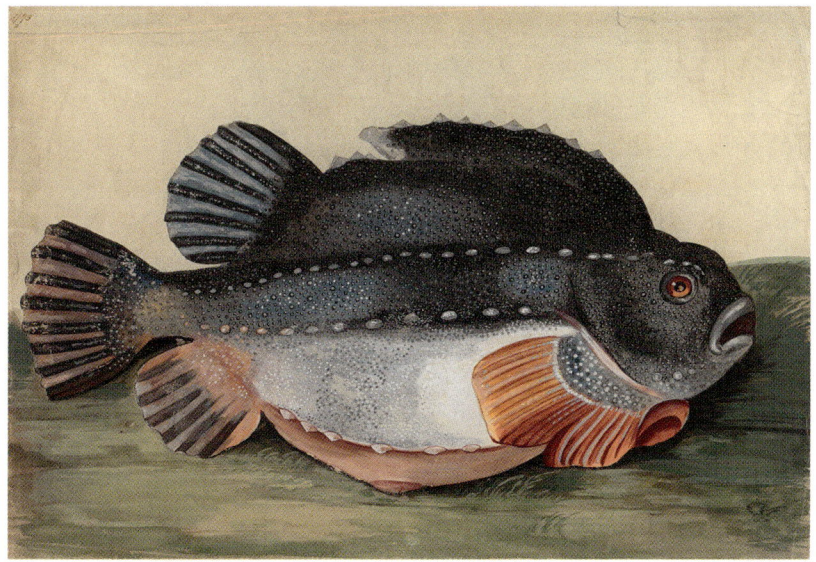

sketching and scientific skills to produce a detailed drawing of the fish's glowing lure, known as an esca, which varies in shape from species to species. A photograph alone of the curled, crumpled specimen would not have done the job.

While there is no beating the atmosphere and detail of modern underwater photography, there remains a limit on what it is possible to capture on film or digital sensor: the fish always turns away, swims off and hides; the whale is simply too enormous to photograph its entire body in detail, all in one go. And it is evident from many of the elegant illustrations I've picked for this book that scientific artworks far exceed photographs of fished, dead specimens. A skilled artist can create an illustration that shows what these animals were like when they were still alive, but without the restrictions of being underwater.

Discovering ocean species goes beyond the academic pursuit of taxonomy, the precise science of assigning names and knowing who is who. It opens our minds to the possibilities of what life can be. Aquatic lives are shaped by rules of biology that are profoundly different to those on land, and as a consequence evolution has filled the ocean with astonishing life forms.

Knowing the species that inhabit the ocean also brings into view how this vast, interconnected living system works. I am an ecologist by training and experience. When I see a fish or an octopus or any other ocean species, I'm caught up not just in how they look but in questions of how they live here, what they eat, where they move and what other intricate components of the ecosystem they depend on. And so, weaving through the pages of this book are ecological ideas, as I've sought images that show marine species in their natural surroundings, interacting with each other and helping to show the inner workings of these wild places.

I have searched for images that show ocean species in such a way as to inherently bestow on them their own rights to exist.

Starfish (or sea stars), from Ernst Haeckel, Kunstformen der Natur
(1904).

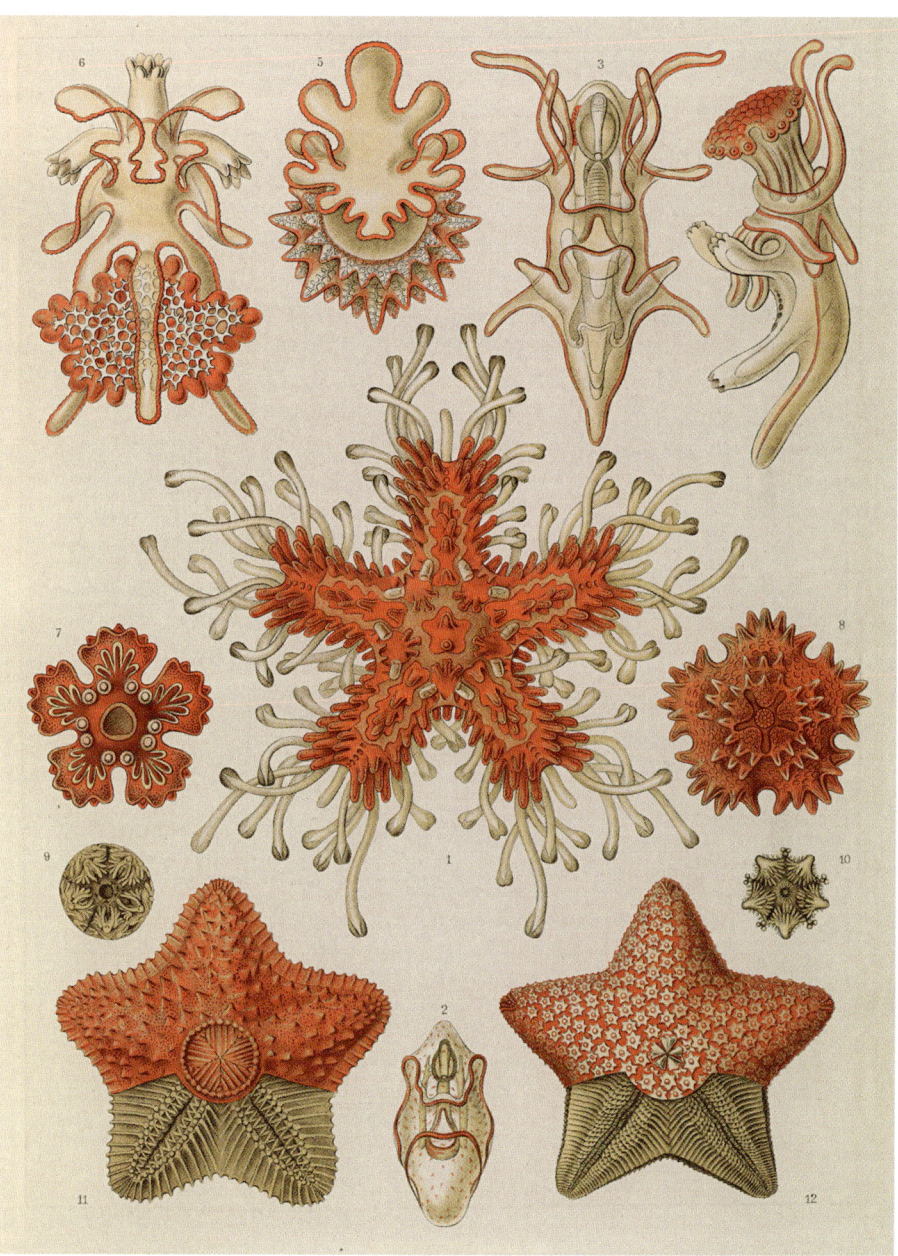

I am purposefully not looking for images that portray fish and other sea life merely as food. That's not to play down the importance of the ocean in providing nutrition and utilitarian goods to humans for millennia, but rather to recognize that the seas are not just there for people to use. Instead, we can look at life in the ocean for no greater purpose than to embrace our own awe and fascination in this huge, hidden portion of the living world.

The works in this book range from fine art and artefacts to ancient rock art, illustrations, ceramics, sculptures and popular culture. Among them are images that were originally intended to be scientific illustrations faithfully capturing the appearance of species, although no less beautiful for their scientific accuracy. Others are detailed and realistic and yet they lie firmly within the realms of fine art, or they are imaginative works that spring from

Below: *Mary Altha Nims,* Shells in Seaweed, *1800s, watercolour.*
Opposite: *George Barbier, 'L'étourdissant petit poisson . . . Robe d'été' ('The stunning little fish . . . Summer dress'), illustration from* Gazette du bon ton *(May 1914).*

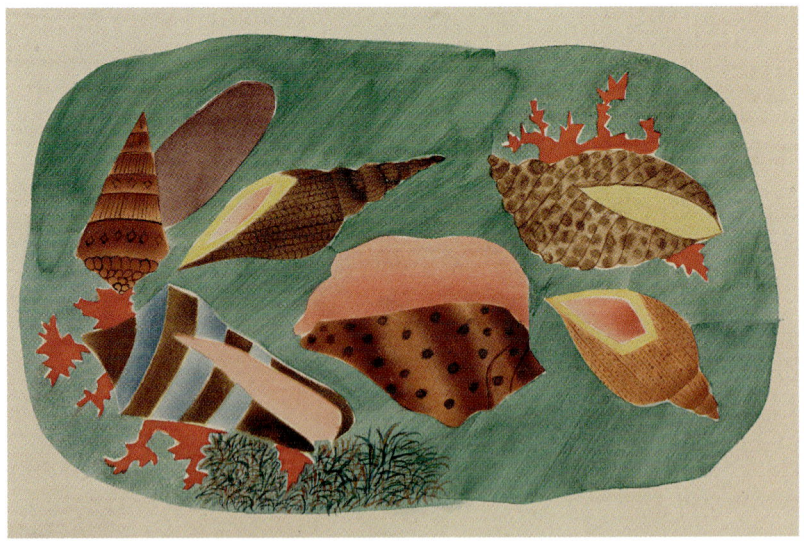

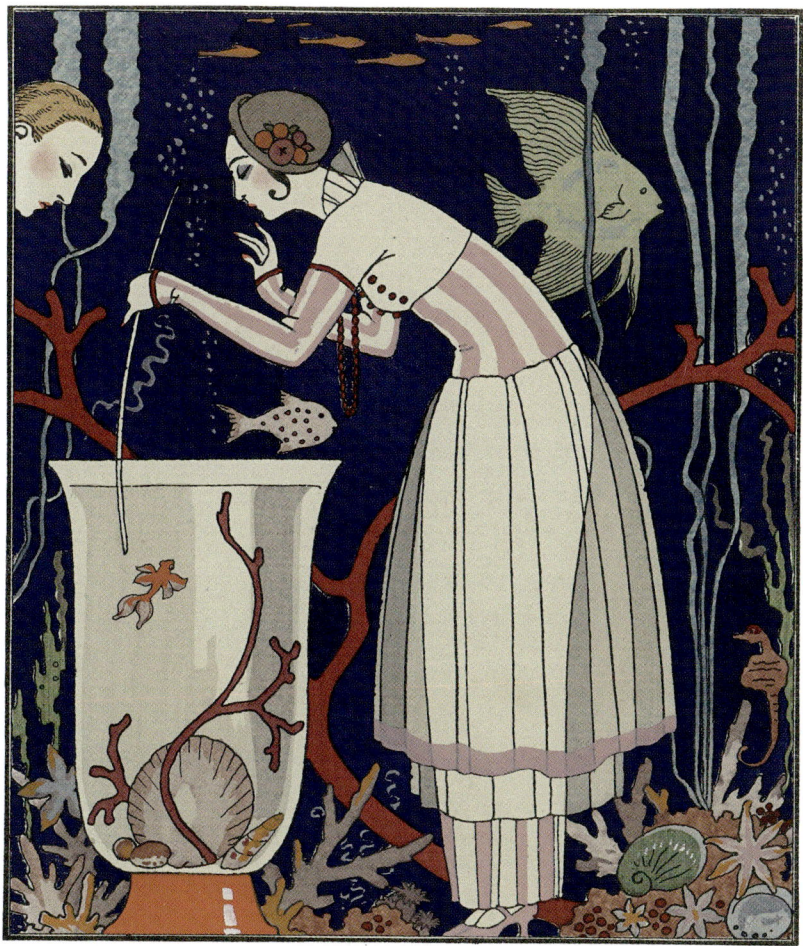

the artists' minds rather than the ocean itself. While there are lines dividing these different types of image, the boundaries are not fixed. Science and art need not be kept firmly apart, and in fact often blend into one another. Works originally intended to be informative and academic have often been embraced by art collectors. Likewise, works of fine art have found their uses within the academic sciences. The works I've selected in the coming pages reflect this fluid perspective on the imagery of ocean life.

This book is by no means intended to be an exhaustive catalogue of ocean life artworks, but a selection of what I think are the most arresting and significant images that I have come across. They span a long period of time, from prehistory to the contemporary era – although I won't be presenting them in strict chronological order – and they come from across the globe.

Two-handled bowl with fish, Italian, 1400–1430, earthenware.

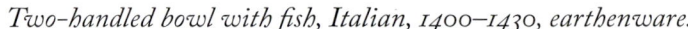

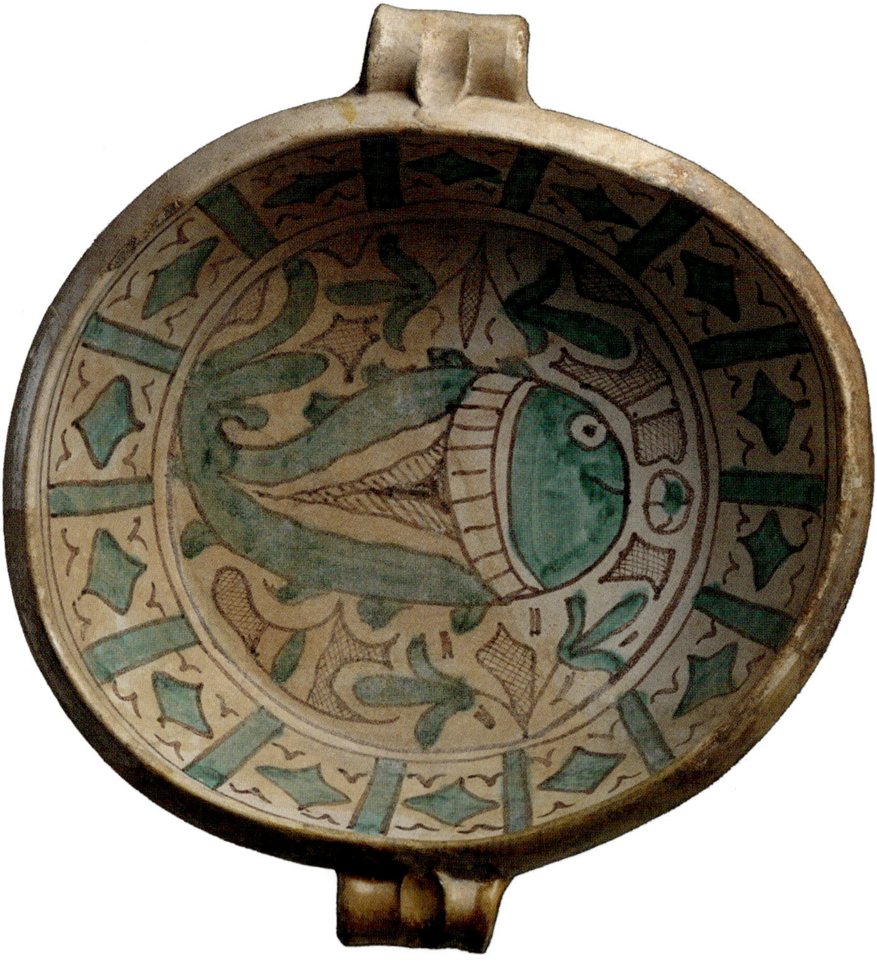

A Tour of the Ocean

This book takes the shape of a journey through the ocean. I could have chosen to go on a tour through the named ocean basins, the Pacific, Atlantic, Indian, Arctic and Southern Oceans, as well as the numerous smaller seas, from the Mediterranean to the Caribbean, Red Sea to Yellow Sea. Instead, we will dip in and out of all these places as we roam the global ocean, a considered term chosen to denote the singular nature of this interconnected body of water that flows around our planet. There is just one ocean.

The journey is divided by chapter into four principle regions of the ocean. We will start at the most familiar part, where the land becomes sea. Walking the coastlines, between the high and low tides, we will encounter the most accessible of the ocean's treasures, those that humans have known about for the longest time and brought into their cultures. Here there are sandy and rocky beaches, wave-swept shores and towering cliffs. This is a transitional zone, a staging area between contrasting living spaces. Much of what we find here is the remains of living wonders from the ocean, which leave tantalizing clues about what's going on beneath the waves.

Next, we enter the shallow seas that fringe the land. These begin where the waterline falls to its lowest point, the places that are never exposed to the air even when the Moon and Sun are aligned and are doing their best to pull the ocean away from the land during the most extreme spring tides. Among scientists, this region also goes by the name of the sunlit or euphotic zone. Everything living in the shallow seas is fed by the sunshine-catching organisms – the minute phytoplankton that look like tiny jewels of so many shapes and types, the algae and plants that nurture underwater gardens and forests, the tiny algae that live inside the bodies of corals and help them build the reefs that fringe tropical coasts. These photosynthetic creatures channel into the food webs energy which then gets passed on from prey to predators and eventually, when those animals die and fall to the seabed, scavengers have their turn and clear up what remains, cycling nutrients back into the water and in turn enabling more food to be made. The sunlight pouring down and abundant nutrients stirred from the land and the seabed make the shallow seas the most productive, biodiverse and species-filled region of the ocean. And with all those creatures comes a tremendous variety of form and colour.

Beyond the horizon we reach the open seas. From far above, perhaps looking out of an aeroplane window, this huge space looks like an empty blue desert. The blue colour of the open seas is a consequence of the way that fractions of sunlight are absorbed by the clear water, in the absence of sediments and dense blooms of colourful plankton. The size of water molecules and the rate at which they vibrate determine how they most strongly absorb the longer wavelengths of light – the reds, oranges and yellows. This leaves the bluer hues to penetrate deeper and scatter from small floating particles and back into the eyes of observers, us, looking down from above.

The lack of any obvious life in the open ocean is a deception caused by distance and a discrepancy in size. There are oceanic giants to spot here, if you're very lucky, including blue whales, the biggest animals ever known to have existed. Though much of the life in the open seas is far smaller and harder to see, it is no less important to the overall functioning and beauty of the ocean. Many are organisms that never know the feeling of a hard surface and spend their lives roaming immense distances across ocean basins. Theirs is a boundless, three-dimensional world.

The final destination in this tour of the ocean will be the hardest part to get to, the places that humans are only just now starting to know and explore. The deep ocean begins at depths where sunlight is no longer strong enough to power photosynthesis. Generally speaking, this is around 200 metres (656 ft) down (although it can be shallower if other factors such as muddy clouds of sediment limit sunlight's reach). This is the start of the twilight

*'The Portuguese man-of-war (*Physalia arethusa*)', drawn from life by Hashime Murayama, 1919.*

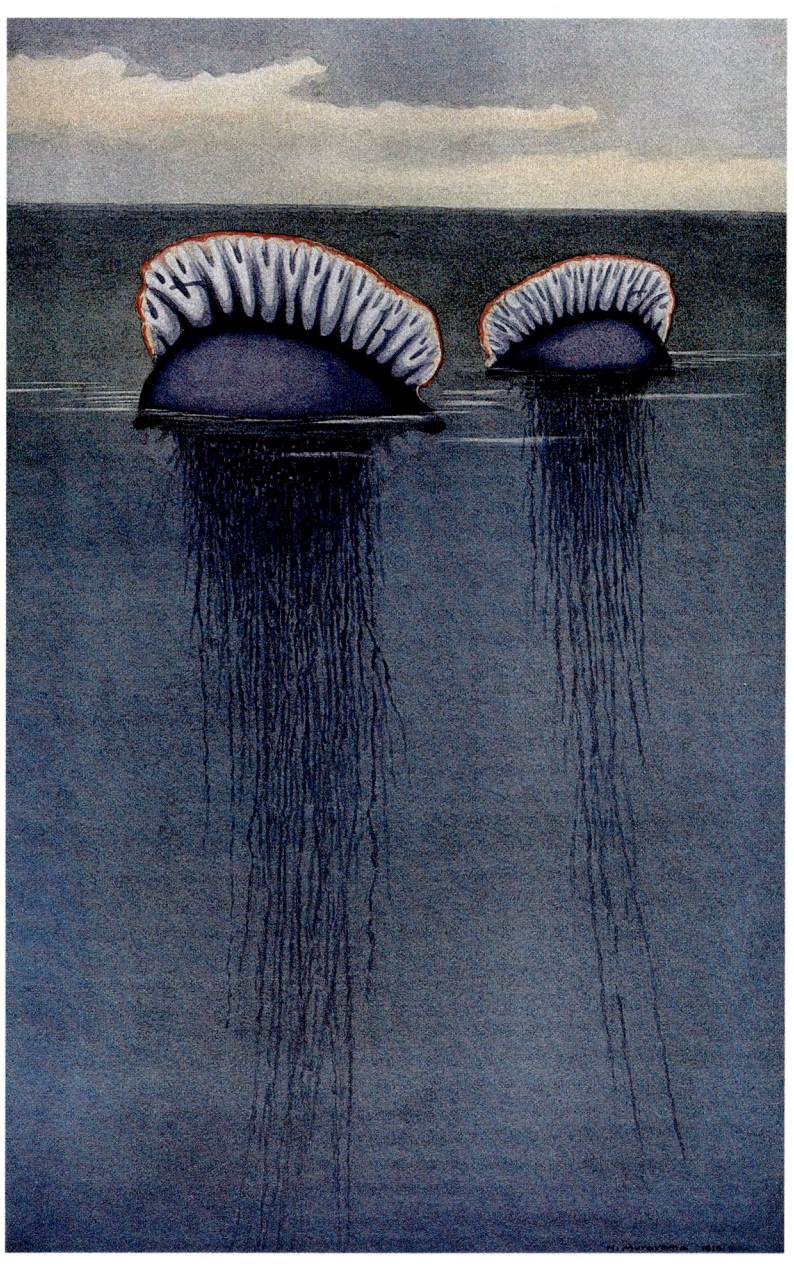

zone, or mesopelagic, a realm where the last scraps of sunlight trickle down. During the daytime, the blue wavelengths fall the deepest, giving this an ambience of an inky blue, cloudless sky a while after sunset. Below this is the midnight zone, or bathypelagic, from 1,000 metres (3,280 ft) down, where there's no sunlight at all, day or night.

On average, the ocean is roughly 4 kilometres (2½ mi.) deep, but in places it plunges deeper into the zone officially known as the abyss, and eventually into the hadal zone (after the ancient Greek underworld, and also known as the hadopelagic), which is made up of giant V-shaped chasms that push into Earth's crust and are frequently more than 10 kilometres (6 mi.) deep.

All the way through the deep ocean, from the twilight zone to the hadal zone, living things face the same set of challenges: it is cold and dark, there's not much food to go around and the pressure of water crushing down gets ever more extreme with increasing depth. In the abyss, the pressure is roughly the equivalent of a fully grown elephant pressing down on every square inch of an animal's body. A human cell would burst at this depth, the membranes cracking under the pressure. And yet countless organisms have evolved to survive and thrive in these relentlessly tough conditions. Their bodies have taken on shapes and structures unlike any we see among terrestrial animals, making them appear the most otherworldly and alien-like, even though, truly, they are earthlings that rule the largest portion of the biggest living space on the planet.

By the end of this journey, we will have seen much of the great breadth of life in the ocean, from the familiar and cherished to the living things that people are still puzzling over and dreaming about. My hope is that the artworks and stories in this book will first and foremost offer a captivating view of the hidden life in the blue parts of the globe, and show that ocean life has been influencing human cultures for millennia and continues to do so today. Now, more than ever, in the swiftly changing times of what has been classed by some in the early twenty-first century as the Anthropocene, it is critical that we find ways to reach our minds into the ocean and feel connections to this distant realm that is so easy to forget and overlook, but which underpins much of what is vibrant and important about the living planet.

ONE
COASTS AND BEACHES

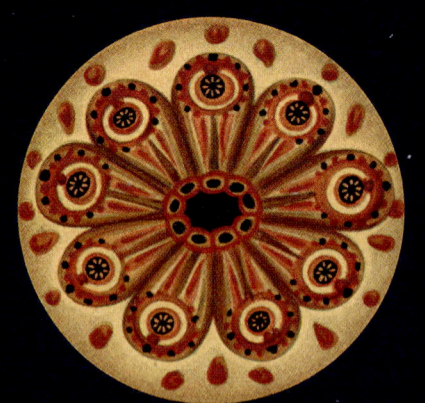

All around the world, between the tidelines at the edges of the sea, live the most accessible of the ocean's inhabitants. Without any specialized equipment, except perhaps a pair of wellington boots, it is possible to walk at low tide on land that was seabed hours earlier and search for creatures that inhabit this liminal space.

Far inland, there are also places that were once seabed much further back in time. This is where countless remains of ancient ocean life lie. These unusual shapes pressed into stone are unlike anything from the terrestrial realm, and people who find them know they are somehow special, imbued with great powers and certainly great meaning.

Centuries ago, when the aristocracy of Europe held lavish banquets, they would set aside a table where food was tested before being served to their guests. It was allegedly a common practice in the Middle Ages and into the Renaissance for people to do away with their enemies by poisoning them, and a banquet was a perfect opportunity to slip arsenic into their food and drink. This was clearly something that scared the nobility, who furnished their tasting tables with state-of-the-art poison detectors.

A few rare examples of these elaborate items of tableware still exist. One dating back to fifteenth-century Germany looks like a small red Christmas tree placed on a hexagonal golden stand with thirteen triangular decorations dangling from its gilt-tipped branches. Each flattened triangle is several centimetres long and has a crenelated gold setting along one side that mirrors the triangle's sharply serrated, blade-like margins. It was said that dipping these amulets into food and drink revealed the presence of poisons either by changing colour or by becoming covered in a damp layer of sweat.

Known in German as *Natternzungenbaum*, meaning 'adder's tongue trees', and in French as *languier*, meaning 'tongue stand', the active components of these ornamental devices likely came from the Mediterranean. For centuries there was a thriving export market in triangular-shaped stones called *glossopetrae* (from Greek words *glossa* and *petra*, respectively meaning 'tongue' and 'stone')

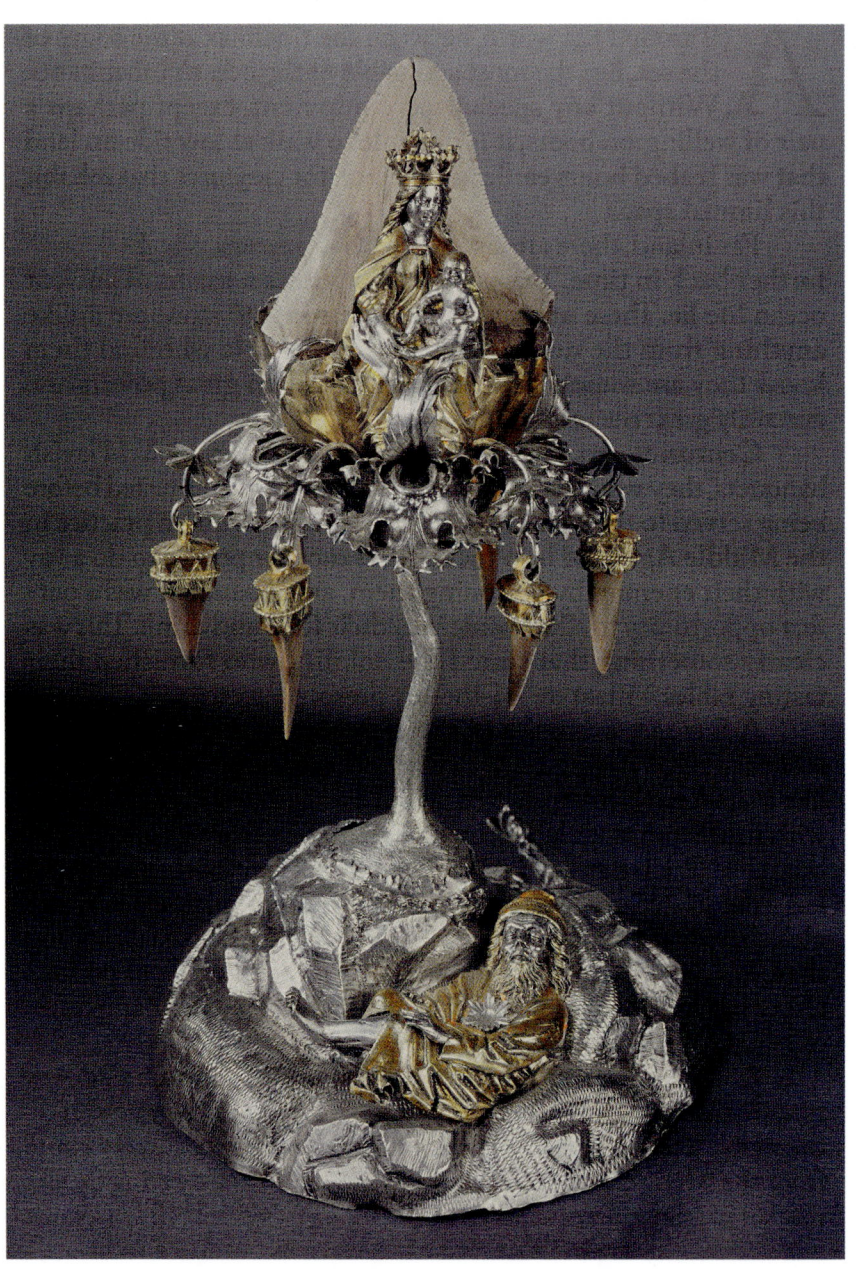

that were dug from the hillsides on the island of Malta. Not only did the trade serve anxious aristocrats who were keen to reassure their dinner guests that the food was safe to eat, but people across Europe wore them as pendants or sewed them into special pockets in their clothing to act as personal amulets and ward off the evil eye. For those who couldn't afford to buy a whole one, pinches of powdered *glossopetrae* were sold as remedies for everything from fevers to bad breath.

Since ancient times the source of *glossopetrae* had been the subject of great debate. Roman writer Pliny the Elder described *glossopetrae* as falling from the skies during dark, moonless nights. A common belief held that these were the petrified tongues of snakes, and so, due to a kind of sympathetic magic, could offer an antidote to snakebites. A biblical story spawned an explanation of how Malta became such a rich source of *glossopetrae*. When St Paul was shipwrecked on Malta and bitten by an adder he sought revenge by turning all the island's snakes to stone.

Other great thinkers were convinced that there must be an alternative explanation for the origin of these curious objects. In the mid-sixteenth century, a physician and naturalist from France, Guillaume Rondelet, studied fish in markets of the coastal city of Montpellier and noticed that *glossopetrae* looked a lot like shark teeth.

A century later, a polymath scientist from Denmark put together a watertight argument for the true origins of tongue stones. Niels Steensen, later known as Nicolas Steno, was brought the head of a great white shark that fishermen had caught off the west coast of Italy. He studied its anatomy in detail, including the multiple rows of large, triangular, serrated teeth. Like Rondelet, Steno became convinced that shark teeth and *glossopetrae* were one and the same thing, although he took his ideas a step further and proposed a theory for how these tongue stones came to be. He

Table ornament in the form of a Tree of Jesse, so-called 'adder tree' or
Natternzungenbaum, *Germany (probably Nuremberg), c. 1500, silver,*
*partly gilded, shark teeth (*glossopetrae*).*

declared that *glossopetrae* were the teeth of sharks that had lived long ago. When the land was covered by sea, these sharpened triangles fell to the sea floor and were buried in layers of sediment, which preserved them and over time changed their chemical composition. Gradually the teeth of ancient sharks had been turned to stone.

It was a wild theory that flew in the face of religious beliefs, suggesting that Earth is far older than the Bible describes and opening the unholy possibility that some forms of life have gone extinct – surely, no wise god would allow his living creations to die out. And yet many *glossopetrae* were far bigger than the teeth seen in any living sharks, hinting that giant sharks had once roamed the ocean.

We now know that many of the medieval Maltese tongue stones came from sharks that lived in the Miocene epoch, between around 20 million and 5 million years ago, including the biggest sharks ever known to exist. Megalodons (*Otodus megalodon*) were around three times larger than modern great white sharks, judging by their 18-centimetre-long (7 in.) teeth. Scientists think that their jaws could have been as much as 3.4 metres (11 ft) wide and their bodies one-and-a-half times the length of a London double-decker bus.

While nobody believes in the curative powers of their fossilized teeth any more, megalodons continue to command great interest and intrigue today. Novelists and Hollywood studios like to stoke myths that there could still be giant sharks alive out there somewhere. Nevertheless, any marine biologist worth their salt will tell you there's zero chance that such colossal sharks have managed to avoid detection by modern scientific equipment. We don't need the possibility of their still existing for megalodons to send thrilling shivers down our spines. Seeing a fossil megalodon tooth or, if you're lucky, holding one, will transport you back to a time in Earth's history when there were indeed huge sharks prowling the seas. Try to imagine the rest of the animal that the massive fang belonged to and what their life was like as they chased after whales and hunted and caught their prey with one of the most powerful bites of any predator.

It is fitting that people revered and treasured the teeth of these magnificent ocean-going animals long before they knew the truth about what they were and where they came from. Show me a shark more worthy of adoration than the spellbinding megalodons. What's more, megalodons and their petrified teeth capture perfectly an enduring relationship between people and ocean life – a connection driven by curiosity and awe.

Ancient Curiosities from the Sea

For millennia, as well as *glossopetrae* many other forms of fossilized sea life have been woven into human lives. Among the most ancient are trilobites, invertebrates that roamed the ocean hundreds of millions of years ago. Ranging from button-sized to almost a metre long, trilobites scuttled across the seabed, drifted through the water column like krill do today or swam around hunting for prey. They went extinct long ago, but can now be seen in their millions, their bodies preserved in rocks all around the world. In a prehistoric cave in France, a fossil trilobite was found perforated on both sides, which hinted to archaeologists that ancient people had used it as an ornament, around 15,000 years ago. Fossil trilobites common in some regions of Spain are known as scorpion stones, because they resemble the segmented bodies of the living arachnids. Legend has it that scorpion stones were used as talismans to protect against scorpion stings, or as an aphrodisiac amulet that, when worn by a woman, compelled her to fall madly in love.

Fossils of enigmatic animals called graptolites were known as Roman stones, because these now extinct creatures left straight-edged impressions in rocks that look, some say, like letters of the Roman alphabet (the word 'graptolite' comes from Greek words meaning 'writing' and 'stone'). Roman stones were used to treat stomach ulcers, as an eye ointment and as wart treatment. Fossilized crab claws were an ancient Roman remedy for stomach troubles and used as curse-repelling amulets.

Sea urchins have existed for hundreds of millions of years and still inhabit the sea floor today, some with long spines, looking like ovoid hedgehogs, some spineless and more flattened that live buried in the sand and are known as sea potatoes. Their ancient, fossilized forms have bewitched people for aeons, chiefly because the five-pointed marks on their rounded bodies make them look like stars that have fallen from the heavens (sea urchins are echinoderms, relatives of the generally five-pointed starfish, also known as sea stars). A reverence for sea urchins dates back at least 400,000 years ago, when a member of the early human species *Homo heidelbergensis* found a flint with a fossil sea urchin and carefully chipped away at it to make an axe with the star shape in the handle. Neanderthals (*Homo neanderthalensis*), our closest human relatives, were also keen collectors of fossil urchins and made them into scrapers and axes. Palaeolithic *Homo sapiens* living in North Africa 35,000 years ago were in the habit of drilling holes in fossil urchins, presumably to wear them as necklaces.

More recently, English folk names for fossil sea urchins include 'shepherd's crowns' and 'fairy loaves'. People placed them on windowsills as good-luck charms to ensure that their bread dough would rise and to stop the milk from going sour. Arranged on the threshold of houses and churches, fossil urchins, it was claimed, would ward off the Devil.

Many people have also buried their dead with fossil urchins. Around 4,000 years ago, in what are now the Chiltern Hills of southern England, a woman was buried holding a young child in her arms and with three hundred fossil sea urchins carefully arranged around them both. At a Bronze Age site in Brittany, France, several large burial mounds have been found that contain no other remains except for a single fossil urchin in each one. Another French burial site further south and made at a similar time contains around 30,000 urchins. These burial practices have a long history. A grave from Anglo-Saxon times in the east of England was found with a woman clasping a fossil urchin in her hands, and another with a woman who had an urchin inside a leather bag hung around her neck. We can only guess at what meanings urchins held, but there was clearly something compelling about sending the deceased into the afterlife, and towards

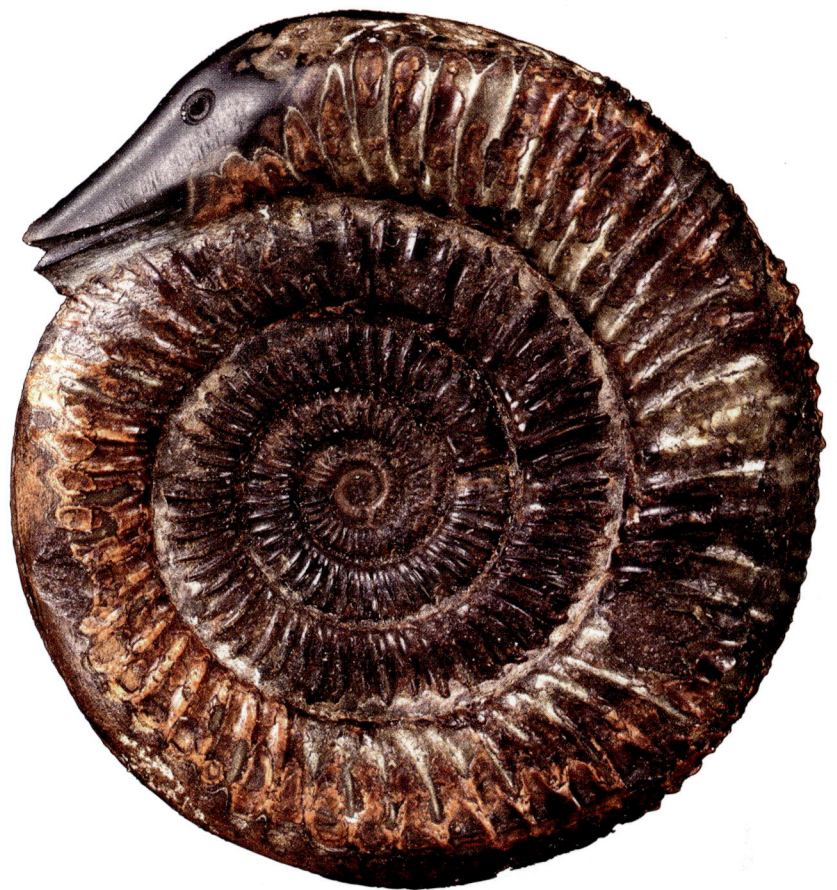

An ammonite fossil, known in English folklore as a snakestone,
with a carved snake head, from Whitby, Yorkshire.

the heavens, accompanied by these earthly incarnations of celestial
bodies.

Stones sculpted into intricate spirals have played a part in
mythologies around the world. A popular story dates back to the
seventh century, when a high-ranking Saxon woman from the
northeast of England, Hilda, rid the grounds of her new abbey
of dangerous snakes by performing a miracle that cut off their
heads and turned them to stone. She then threw them off the

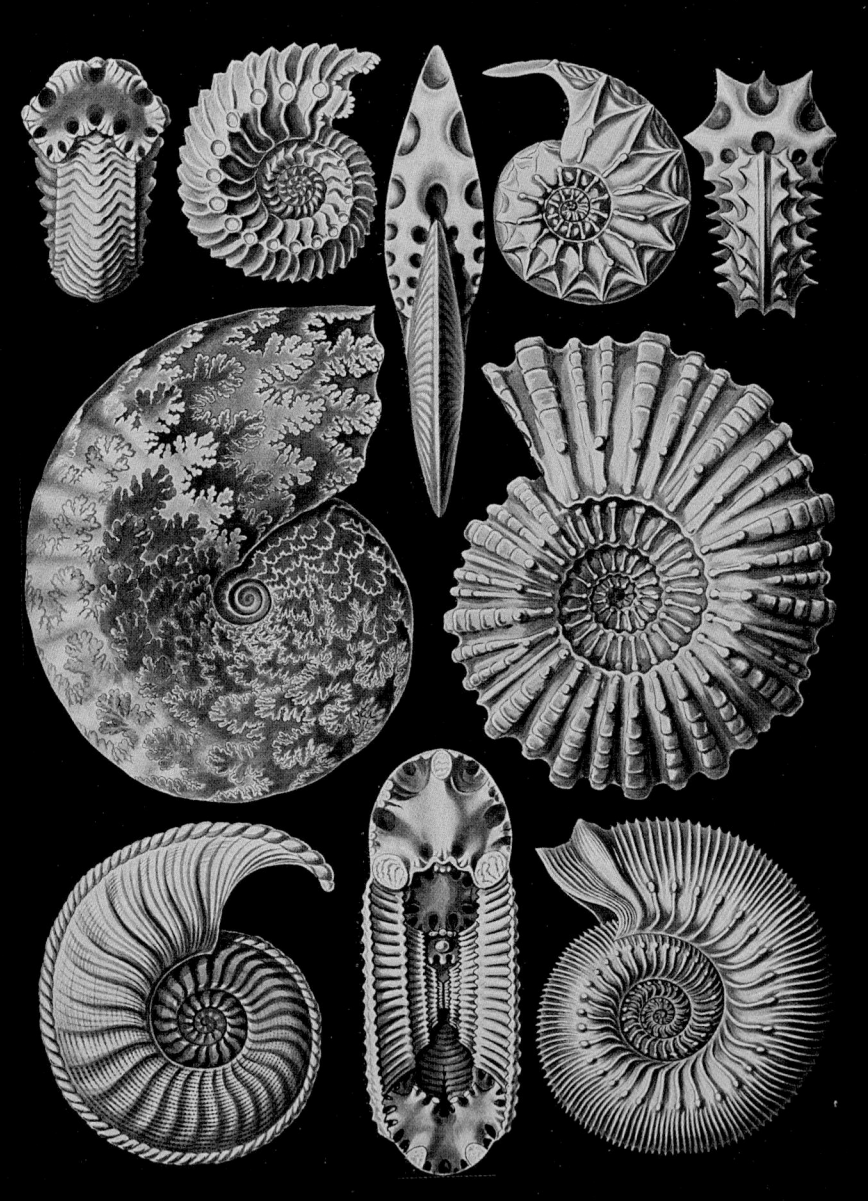

cliffs, where the petrified coiled snakes can still be seen today. Hilda was made a saint and centuries later her story sparked a Victorian craze in snakestones. Entrepreneurs collected the coiled stones from the cliffs around Whitby in Yorkshire and had snake heads carved onto them. Even though Hilda was supposed to have chopped off the snakes' heads, these souvenirs were a hit with tourists, who were flocking to the coast on the newly opened railways.

It was around this time that Victorian scientists were beginning to understand much more about the true nature of snakestones and other mysteriously shaped rocks. Fossil hunters were digging up all manner of ancient sea creatures, such as the giant marine reptiles found by Mary Anning on the Dorset coast that now line the halls of London's Natural History Museum, like a display of so many long-lost sea monsters. Likewise, the abundant spiralling stones were not made by any living species but by a group of extinct molluscs, relatives of squid and octopuses, called ammonites. The closest living analogy to ammonites are the chambered nautiluses that live in the depths of the ocean and rise towards the surface at night to scavenge and grab food with dozens of sticky tentacles. Like trilobites, ammonites were readily fossilized thanks to their tough shells, but fossilized remains of their soft bodies have never been discovered so we can't be sure what the living animals looked like. Their spiralling shells were divided into interior chambers and the main part of their bodies would have occupied the outermost chamber, as they do in nautiluses today. The inner chambers were filled with gas, making the shell more buoyant and saving the ammonites energy while they swam through the ocean by jet propulsion, squirting water through a bendy siphon tube. Presumably, ammonites had tentacles sticking out of their shells to catch prey, maybe noodle-like and similar to those of nautiluses, or maybe extendable sucker-tipped clubs like cuttlefish and squid.

Fossil ammonites, from Ernst Haeckel, Kunstformen der Natur *(1904).*

Ammonites were undoubtedly prolific and immensely successful in their time, and their fossils can be found worldwide, where people have put them to many uses. Geologists have used them as markers of time to estimate the age of rocks, because different species with distinctive shells existed at different times in the distant past. More whimsical uses include the ancient Romans' placing a golden ammonite under their pillow, believing that it would let them dream the future. In Germany, farmers called them dragonstones and believed that putting one in an empty pale would encourage their cows to produce more milk. In Scotland they were called crampstones; soaking an ammonite in water for a few hours yielded a liquid that was used to wash ailing cows and horses to restore their health.

There are no living ammonites to be found anywhere in the ocean today. The last of this lineage of animals went extinct 66 million years ago in the same mass extinction that wiped out the dinosaurs, together with 75 per cent of all Earth's animals. When a giant asteroid, Chicxulub, slammed into the planet, it threw dust into the skies, blocking the Sun and causing ecosystems to collapse because plants couldn't grow – within a few weeks, global photosynthesis shut down. At the time there were immense volcanic eruptions that sent up plumes of sulphur dioxide which then fell as acid rain, lowering the ocean's pH and wiping out the plankton that many other ocean-going animals, likely including young ammonites, depended on for food.

Palaeontologists continue to study fossil ammonites and unpick more of the untold secrets of the past. And these ancient cephalopods are still popular among fossil collectors, with many beautiful specimens coming from the Indian Ocean island of Madagascar. Each ammonite fossil is a memory trapped in stone of an ancient living world and a reminder that ocean life is ever-changing and fragile. Even animals that once ruled beneath the waves in tremendous abundance have met their end, and we must not forget that it can easily happen again, especially now humans are so badly mistreating the ocean and ocean life.

Spellbinding Seashells

The most common natural objects to find on the shoreline are seashells. Cast-off exoskeletons of molluscs that were recently alive, and not long dead like their ancient cousins the ammonites, wash up on coasts and beaches everywhere. Winkles, whelks, top shells, tower shells, cowries, clams, conches, cockles, mussels and many other living varieties of shell-making mollusc spend their lives hidden down on the seabed, or in the places between the tides that switch from being sea to land then back to sea again. These soft-bodied animals make a single shell, or in the case of bivalves a pair that press tightly together, which they live in throughout their lives and expand incrementally as their bodies grow bigger inside. The raw materials for making shells are calcium and carbonate ions dissolved in seawater, which the molluscs absorb, then lay down in layers of limestone.

A shell is a multipurpose structure for molluscs – it is primarily a protective home that is resistant to attackers and to dehydration for the intertidal dwellers that are exposed daily to the air and sun; it is a point for muscle attachment much like the internal skeletons of vertebrates; it can be a tool to assist in hunting; and it can be a window to let the sunshine pour in so that tiny algae can grow inside and provide crops of food for the living mollusc.

In a similar way to the molluscs that make them, humans find many uses for seashells. At their simplest, shells are pleasing objects to pick up and hold, generally just the right size to slip in a pocket as a souvenir from a day at a beach. Some people take their shell collecting more seriously, from amateur naturalists keeping records of what they've seen to taxonomists searching for previously unknown species and helping to understand and appreciate Earth's biodiversity – there are tens of thousands of named species of shell-making mollusc and countless thousands more no doubt await discovery.

Shell crafters have gathered shells to make and sell decorative items such as sailor's valentines, which were popular in the nineteenth century. After months at sea, mariners working on

North American commercial whaling ships would bring back these mementos for their sweethearts. These trinkets were made from hundreds of tiny shells glued into patterns on cotton batting and often encased in a pair of octagonal wooden boxes. Unlike the scrimshaw designs scratched into whale bones and teeth, the sailors didn't while away the long, dark nights gluing shells into pretty patterns. Instead, towards the end of their journeys, whaling ships would often call into the Caribbean island of Barbados, where local women were shell artisans creating souvenirs for visitors.

Earlier than this, sailors on seventeenth- and eighteenth-century merchant vessels returning from Indonesia brought back a different kind of molluscan memento. Nautilus cups were made from the carved shells of chambered nautiluses and presented in ornate golden mounts, turning them into the hulls of sailing ships, or the bodies of birds and sea monsters. Often, the outer layer of the nautilus shell was scraped away to reveal the iridescent mother-of-pearl, or nacre, underneath. This under-layer of their shells gives the chambered nautilus their alternative name, the pearly nautilus.

It's common for molluscs to line their shells in nacre. This material isn't produced in order to look attractive but to be smooth and incredibly strong. Nacre is a form of calcium carbonate that is highly crack-proof, helping molluscs to resist the crushing claws and crunching jaws of predators. Molluscs use the same material to make pearls. If a parasite, a piece of grit or an irritating grain of sand invades their body, many molluscs deal with the offending particle and make it inert and safe by smearing it in layers of nacre, which build up to form the gleaming orbs that people have obsessed over for centuries. Nacre-lined shells also have their uses in the human world. Button makers stamp circles from pearl mussel shells. And the swirling, lustrous nacre from abalone shells is used in decorative inlays in furniture, jewellery and musical instruments.

Monumental seashell-gathering efforts are evident in the shell grottos built over the past few centuries by the British upper

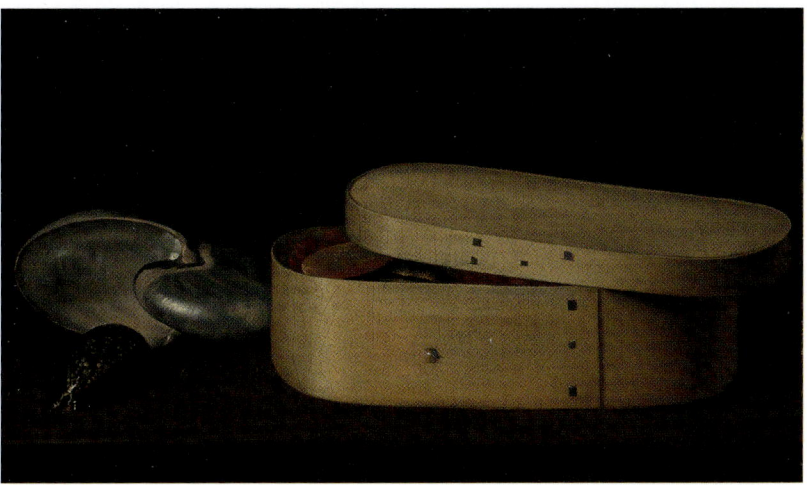

Sebastian Stoskopff, Still-Life with Shells and a Chip-Wood Box, *late 1620s, oil on canvas.*

classes. Many country mansions had a folly in the garden with walls encrusted in shells. Some were intended to mimic natural sea caves and had water features running and dripping through them, and some had more ornate and decorative mosaics, a rendition of Neptune's underwater palace. In Margate on the Kent coast, a 20-metre-long (66 ft) underground tunnel is estimated to have more than 4 million seashells pressed into the walls, and nobody is sure who put them there or why.

A long way further back in time, shells were being fashioned into objects that played significant roles in the lives of very early humans. On the Indonesian island of Java, archaeologists have uncovered a mussel shell with simple zigzag patterns that were carefully etched into its surface an astonishing 430,000 years ago. The location and age of the etched shell suggest that this was the work of *Homo erectus.* The mussel was drilled into with a sharp tool, perhaps a shark's tooth. Archaeologists dare not suggest a meaning and purpose for the patterns but the finding undoubtedly shows that the ancient ancestors of *Homo sapiens* were dextrous and thoughtful.

It is well known that shells provided materials for early humans to fashion ornaments for their bodies. Remains of shell beads have been found across the African continent, in Morocco, Algeria and South Africa, many that were made at least 100,000 years ago. The oldest known examples, a cluster of 33 shells from a cave in Morocco, were rubbed with ochre pigment, drilled with holes and worn together on a string at least 142,000 years ago. Bearing in mind that *Homo sapiens* likely evolved as a distinct species between 200,000 and 300,000 years ago, crafting jewellery was a remarkably early milestone in the development of modern human behaviours and identity. There must have been some significance to this practice, perhaps a sign of status or belonging, and it surely indicates that people were thinking about the world around them and considering their relationship with it.

As humans migrated across the globe, they continued producing jewellery from shells. In Australia, the earliest evidence of this comes from a collection of 22 beads made from cone shells found in Mandu Mandu Creek rock shelter, near Exmouth on the coast of Western Australia, dating back 32,000 years. In Jerimalai cave on the island of Timor-Leste, a site known to be occupied around 42,000 years ago, archaeologists have found pieces of nautilus shell that were smoothed down to reveal the shining nacre layer underneath and stained with red ochre pigment.

Seashell jewellery and other shell decorations have appeared in just about every human culture with access to the coasts and many located far inland. Many have been preserved and found in graves as it has been a common practice to bury the dead surrounded by shell objects, adorned in shell jewellery and dressed in clothing with shells sewn into them. Around 12,000 years ago in a rock shelter on what is now the Indonesian island of Alor, a fisherwoman was buried with five crescent-shaped fish hooks carved from shells tucked under her chin. Near the southern French city of Avignon, a Stone Age man was buried 7,000 years ago wearing a tunic neatly embroidered with hundreds of seashells. Several centuries later, a great city of the dead was built in what is now Bulgaria, containing three hundred graves, many of them filled with treasure. The Varna necropolis contains the oldest known buried gold in Europe, as well as items of jewellery intricately sculpted from seashells. People have also routinely sent their dead into the afterlife accompanied by intact seashells, even in sites a long way from the ocean. The Scythians were a group of nomadic warriors in the first millennium BC who roamed the grasslands of Central Asia on horseback and built burial mounds decorated with cowrie shells.

The cowrie is one of many shell types that have been used as an early form of currency. In China, the classical character for money incorporates a pictograph of a cowrie. They were used as currency thousands of years ago, and when there weren't enough real shells available, replicas were made from metal and bone. Cowrie shells became a central currency in the European slave trade. In all, more than 30 billion cowries were taken from the Maldives archipelago in the Indian Ocean and used as cheap ballast to steady the European merchant ships that filled their holds

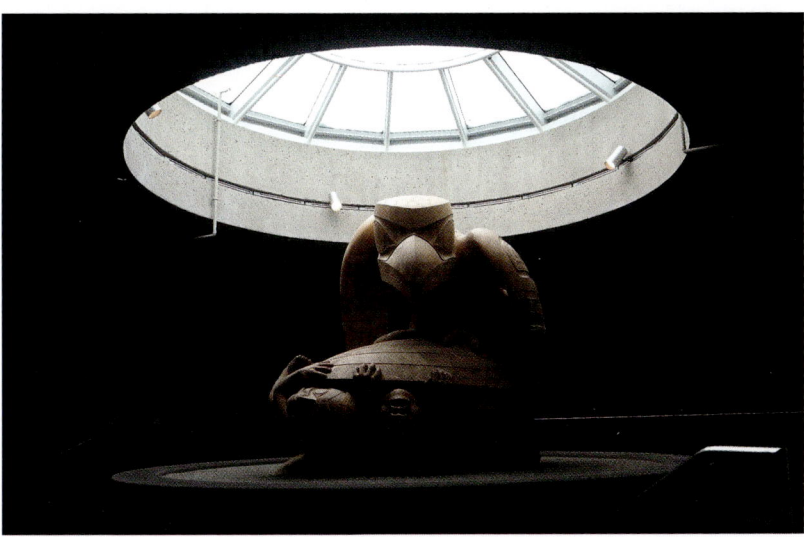

Iljuwas Bill Reid, The Raven and the First Men, *1980, yellow cedar, laminated and carved,* UBC *Museum of Anthropology, Vancouver.*

with spices, fine china and silks. The shells were then repacked in Europe and sent to West Africa, where they were exchanged for human lives.

By their very nature, seashells lend themselves to being practical and symbolic objects for humans to make use of. They provide durable, workable raw materials; their white colour represents life and death in many cultures; their great abundance is a potent sign of fecundity; and finding shells washed up on beaches has compelled people to cast their minds into the hidden ocean depths, to ponder the origins of life, and thus shells have become entwined in many creation stories, old and new. On Canada's Pacific Coast, people of the Haida Nation tell a creation legend in which Raven finds a clam-shell on a beach and releases from inside it the first Haida men. Sandro Botticelli's *The Birth of Venus* depicts the Roman goddess of love newly emerged on her scallop shell.

Artful Shells

Beyond their use as physical objects, the elegance of molluscan shells, their shapes and patterns, have long held enduring fascination for artists, architects and designers. Many shell admirers have focused on the graceful spirals that all seashells are based on – some more obviously than others. A nautilus shell cut in two reveals the inner spiral that runs through its middle and which obeys the basic mathematical rules of the logarithmic spiral: it expands by the same amount for each 360-degree turn (the shell-making molluscs are not aware of their arithmetical expertise – they just happen to create these shapes as they gradually expand their homes and keep them habitable). Meanwhile, a scallop shell is also at heart a logarithmic spiral, only one that expands very swiftly so that rather than turning around on itself as snails do, it flares into a flattened half-shell.

Illustrations of murex shells by George Brettingham Sowerby II, from Lovell Augustus Reeve, Conchologia iconica, *vol. III (1845).*

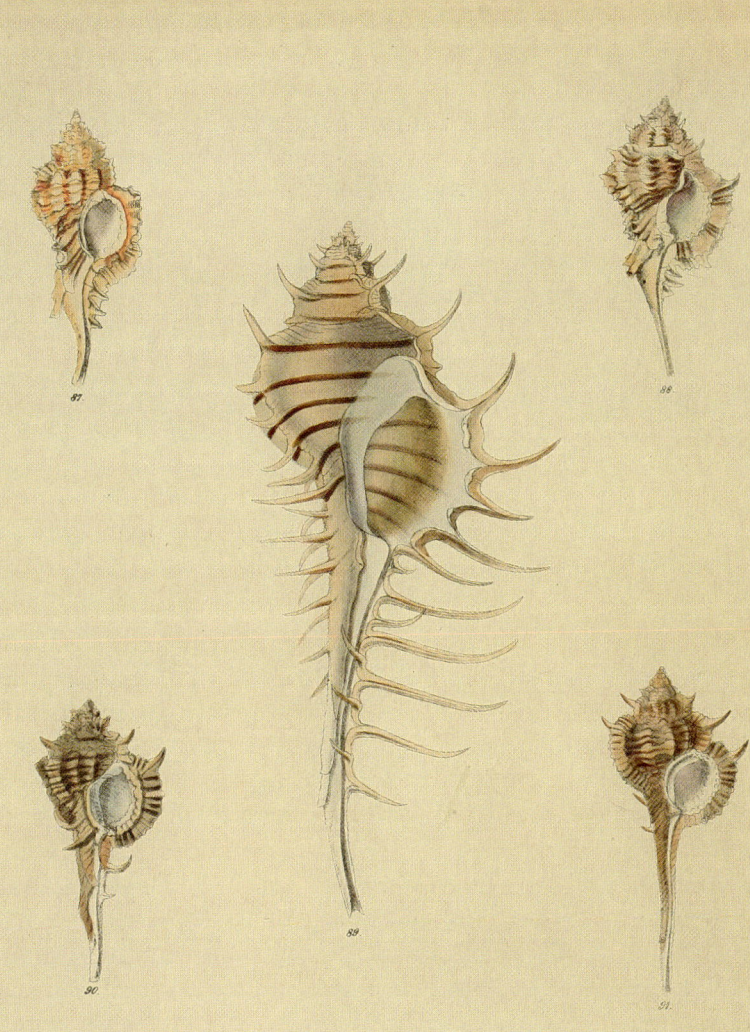

Art and science came together in pre-photography times, when many scientific illustrators aimed to accurately depict the immense variety of forms and the inherent beauty of seashells. Undoubtedly, the most ambitious book of shells is *Conchologia iconica*. The twenty volumes took 35 years to complete, between 1843 and 1878, by a pair of British conchologists and illustrators: Lovell Augustus Reeve and then, following his death, George Brettingham Sowerby II. Their monumental work comprises full-size, lifelike colour lithographs of around 27,000 shell species from all around the world. It was a pre-eminent guide for shell collectors and natural historians, at a time when there was considerable overlap between those two groups of people.

Many of the shells illustrated in these immense volumes belonged to a British conchologist, Hugh Cuming, who had devoted much of his life to travelling the world on self-funded expeditions to gather shells from remote locations. He commissioned a wooden schooner called *Discoverer*, which was likely the first vessel built expressly for the purpose of making natural-history collections, with plenty of room for boxes upon boxes of shells. Cuming sailed across the eastern Pacific, from the coast of South America via Easter Island to Tahiti and back. He collected shells in Central America and Galápagos. He also went to the Philippines before this rich hotspot of biodiversity became a popular destination for naturalists. When ill health put an end to his exotic travels, he returned to London with an unrivalled shell collection, which he then augmented with specimens bought in the auction houses of Europe. At the end of his life, in 1865, Cuming had amassed around 83,000 shells, which were sold to the zoological collections of the British Museum, now the Natural History Museum in London. Cuming's shells can still be found dotted through the museum's collection of more than 8 million shells and illustrated in many pages of *Conchologia iconica*.

A Shell Compendium

Smaller collections of shells have long been a common feature in still-life paintings. Dutch painter Henricus Franciscus Wiertz picked a selection for *Shells and Marine Plants* (1809) that encapsulates much of humanity's preoccupations with seashells. Overlooking what appears to be a moody grey-green sea, arranged on a cliff's edge, beneath a rocky pinnacle topped in the lacy webs of fan corals, is a pile of exotic shells.

The eye is drawn to the largest shell on display, a great triton (*Charonia tritonis*), with scalloped, banded patterns encircling its elongated spire. This is one of the world's largest species of snail (otherwise known as gastropods), with shells that can be 50 centimetres or close to 2 feet long. Native to the shallow tropical seas of the Indian and Pacific Oceans, living tritons prowl coral reefs at night seeking the scent of prey, including enormous, spiky crown of thorns starfish. The triton engulfs the target in a big, muscly foot, then uses paralysing secretions to render the starfish helpless while slowly digesting its innards. Tritons are named after a comparably truculent character, the mythical Greek demigod of the same name who blew on his shell trumpet to control the seas and vanquish enemies.

In the real world, tritons and other large shells, including various species of conch, have been fashioned into musical instruments. *Horagai* trumpets are used in Buddhist temples in Japan, *dung-dkar* in Tibet and *pūtātara* trumpets are made by the Māori of Aotearoa (New Zealand). Shell trumpets are blown at the funerals of Fijian chiefs, and in Haiti they are played to summon water spirits.

The physics of sound production in these molluscan instruments is much the same as brass instruments: the large, hollow bell of the shell acts as a resonating chamber, like a trumpet or trombone, amplifying and reverberating the sounds generated by buzzing lips pressed to the open tip. For similar reasons, the sound of the sea apparently becomes trapped inside large seashells and can be heard when the shells are held to an ear. Low-level ambient

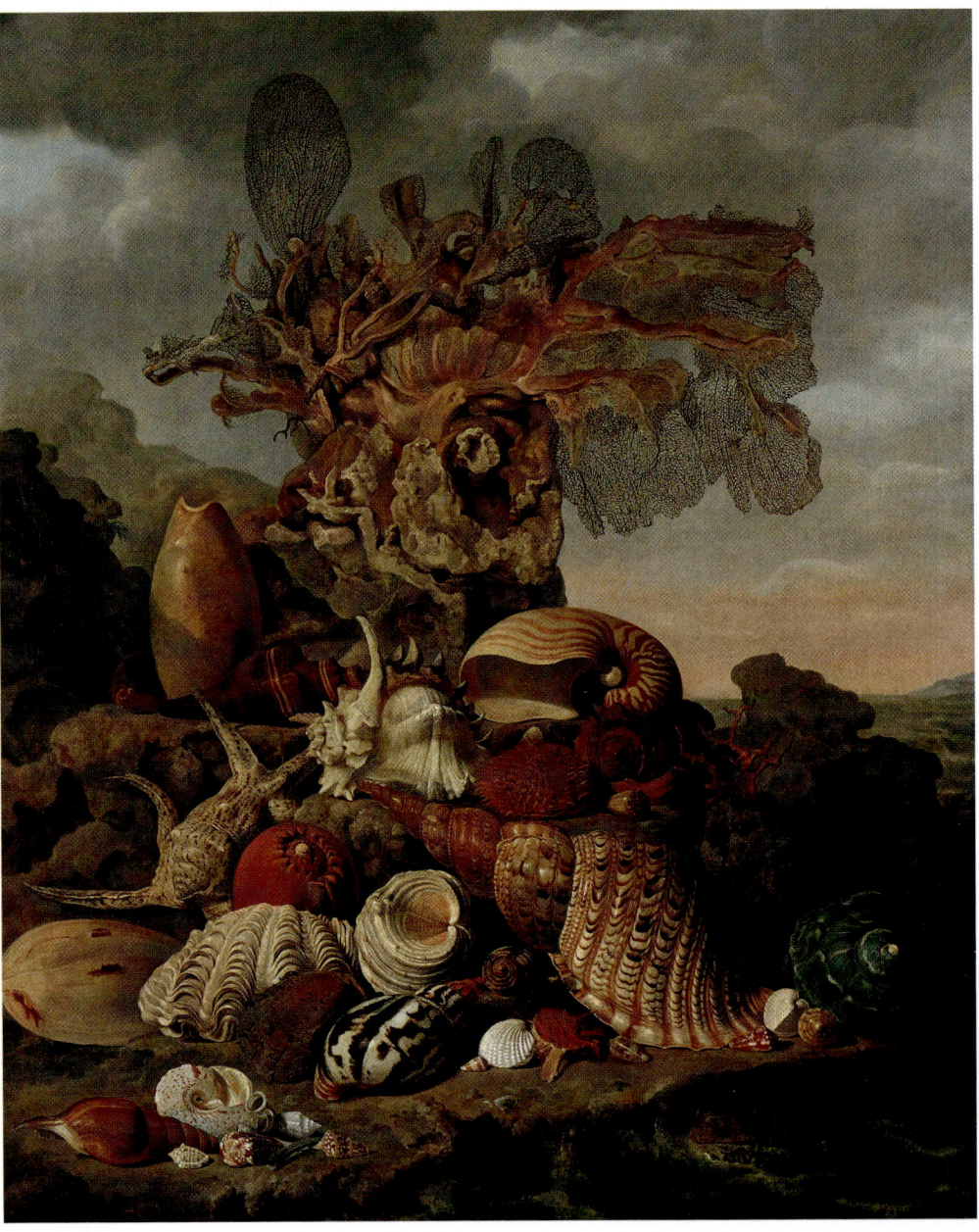

noises, such as a gentle breeze or the whoosh of blood through your ears, enter the shell and bounce around inside, mimicking the sounds of waves. The simplest shell trumpets can be made just by cutting the tip off. More ornate versions have mouthpieces added, and holes cut in the side to produce different musical notes.

Just above the triton shell in Wiertz's painting is a queen conch (*Aliger gigas*), a large, white, fluted shell with a prong sticking upwards. This species originates in the tropical waters of the western Atlantic, between Brazil and Bermuda. The large snails are herbivores that roam green shallow seagrass beds. They've played an important role in many cultures. Conch meat has been eaten for centuries, as is evident from immense piles of their empty shells in ancient middens across the region. Indigenous people in the Caribbean have used conch shells to fashion tools and make trumpets. Aztec carvings depict warriors and gods brandishing conch trumpets, known as *quiquiztli*. Mayan carvings show fighters holding conch shells like boxing gloves, perhaps protecting their hands. In the seventeenth and eighteenth centuries, European sailors brought back conch shells and they became popular as household decorations and to carve the relief images of cameos, often portraits of important people.

Pearly pink conch shells have been widely used as inlays in jewellery and ornaments, sometimes alongside the ruby red shells of thorny oysters, several of which can be spotted in Wiertz's painting. These bivalves from the genus *Spondylus* live all around the world in warm and tropical seas. They aren't true oysters (for that, they would need to belong to the Ostreidae family) but they do share common characters, such as cementing their twinned shells firmly to rocks. Like many bivalves, thorny oysters are filter feeders, drawing seawater into their bodies through a siphon tube and sifting out small particles of food with their gills. Their external spines protect against attacks from predators and also encourage sponges and seaweeds to settle and grow on them, thereby not only

Henricus Franciscus Wiertz, Shells and Marine Plants, *1809, oil on canvas.*

providing a camouflage disguise but further deterring predators with foul-tasting, toxic chemicals.

Thorny oyster shells have been important in cultures on both sides of the Atlantic for millennia. Neolithic Europeans carved thorny oyster shells from the Aegean Sea into bracelets and beads that were traded across the continent. The shells were also the centre of trades across Andean and Mesoamerican societies in pre-Columbian times. These powerful symbols of the sea held immense value. Whole shells were placed in graves and they were used to make masks and mosaics. The mummified body of a twelve-year-old girl from the Incan Empire was found near the peak of a high volcano in Peru. She was one of several children led up the volcano and offered as sacrifices, who were left with objects including a human figurine wearing a red crown fashioned from thorny oyster shell. Centuries before the rise of the Incas, the Moche people thrived on the northern coast of Peru. They carved thorny oysters into beads, called *chaquira*, which were embroidered onto clothing as a form of body armour worn by warriors. And moche artisans created representations of thorny oysters in their pottery.

A trio of large shells from the *Melo* genus appear in Wiertz's still-life. Two are white and one, which is red, is nestled in the middle of the shell pile and shows its spiralling crown of short spikes. What you can't see are the unusually wide apertures of these shells. Commonly known as bailer shells, for centuries seafaring people have used empty *Melo* shells as scoops to bail water from their vessels. When these sea snails were alive, a large head and muscly foot covered in black and white zebra stripes would have stuck out of the large openings of their shells.

A prominent pair of specimens in the painting are the pale, fluted shells of giant clams (one is seen from the side; the other is end-on, forming a heart shape). The genus *Tridacna* contains some of the biggest, heaviest seashells ever known to have existed. They can grow to more than a metre across, weigh a quarter of a tonne and live for a century or more. Living giant clams smile upwards from the seabed with colourful, crimped lips that are in fact another part of this headless animal's body, called the mantle, which extends between their open twinned shells. Minute symbiotic algae living inside them help to feed the clams with sugars.

Various species of *Tridacna* clams have been involved in human lives since very early times. Findings of fossilized remains of *Tridacna squamosina* have hinted that humans hunted them nearly to extinction in the Red Sea around 125,000 years ago, making them one of the earliest cases of the overexploitation of a wild species. More recently, giant clam-shells have become highly sought after and a global trade is proliferating for the shells as a replacement material for elephant ivory. Ornaments and beads carved from giant clam-shells look a lot like ivory and, like elephants, clams are protected in many countries because of concerns over their disappearance in the wild. Similarly, the large green shell, furthest to the right in Wiertz's painting, is a turban shell (*Turbo marmoratus*), an overharvested species that has been depleted in parts of its tropical range for its thick gleaming nacre, which people make into jewellery and buttons.

There are concerns for the future of another of the large shells that Wiertz depicted in his collection. A chambered nautilus shell lies on its side, its tawny tiger stripes still intact and not scraped away to show the nacre underneath. International trade in nautilus shells poses a threat to their survival in some places, in particular the Philippines, where every year tens of thousands of animals are caught in deep-sea traps and killed for their shells.

Scientists continue to learn more about the mysterious lives of these cephalopods. Until recently there were thought to be six species in the *Nautilus* genus, each with a slightly different shell and living in particular regions of the ocean. In 2023 three more species were identified off the Pacific islands of Fiji, American Samoa and Vanuatu. Nautilus experts think there could be many more species still to be discovered living around giant underwater mountains, called seamounts. With their gas-filled interiors, nautiluses can't swim deeper than 800 metres (2,625 ft) because their shells aren't strong enough to resist the crushing pressure and would implode in such conditions. Because they don't swim through open water but prefer to hug the seabed, nautiluses are confined to the flanks of their isolated seamounts. Dotted across the ocean there are thousands of unexplored seamounts that can be several miles tall without reaching the sea surface. It is likely that many of these deep-sea peaks have their own resident, endemic nautilus species swimming around and around.

Several early twentieth-century artists turned their gaze on the chambered nautilus. Julie de Graag (also known as Anna Julia de Graag) was a printmaker and painter from the Netherlands. In 1921 she made a woodcut, titled *Shell*, depicting an intact nautilus shell lying on its side. The clean lines portray the solid feeling of this three-dimensional shell and the aperture faces the viewer at such an angle that it invites them to contemplate what may lie hidden inside the inner chambers, around the curve out of sight.

Below: *Julie de Graag*, Nautilus Shell, *1921, print.* Opposite: *Edward Weston*, Nautilus Shell, *1927, gelatin silver print.*

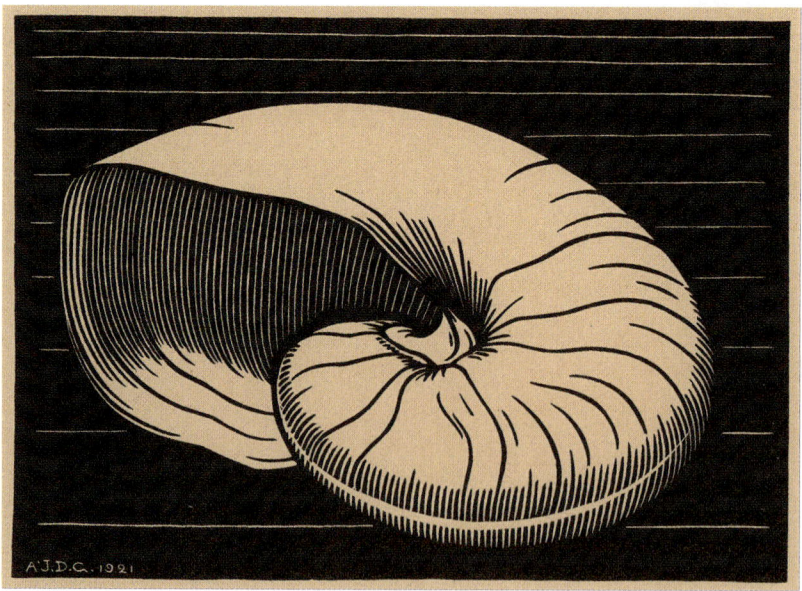

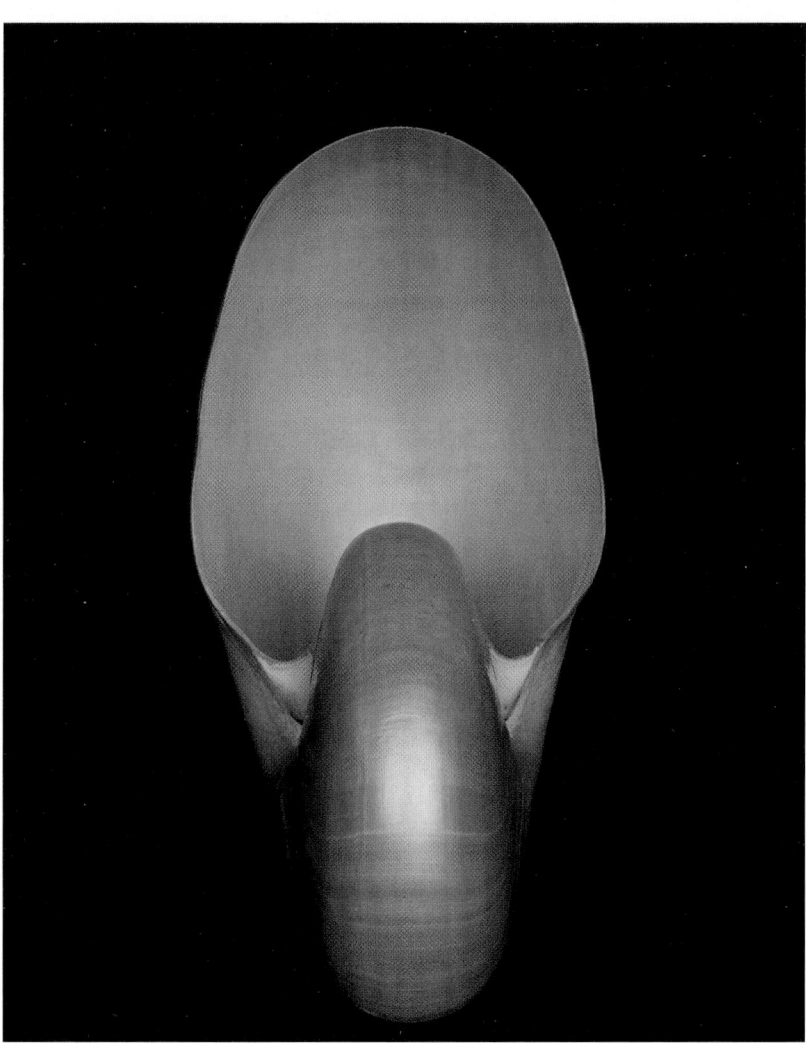

In 1927 the American photographer Edward Weston produced his most famous work, which would later became one of the most famous and expensive photographs ever made. He portrayed a nautilus shell from an unconventional perspective. The shell floats upright against a dark background, and the empty, outermost chamber faces directly ahead. If it had still been occupied by a living animal, we would be staring straight into the eyes of the nautilus.

Weston was inspired to photograph shells by the works of the Canadian painter Henrietta Shore, who like him lived in California. He later wrote, 'I never saw a chambered nautilus before. If I had, my response would have been immediate.' He borrowed several nautilus shells from Shore's collection and set about photographing them in his studio in Glendale, propped up on the end of an oil drum. In order to create the gleaming white image of the shell against the dark surroundings, Weston used a large-format camera set to a very small aperture (f/64), which widens the depth of field and brings both background and foreground into focus. Using film with very low light sensitivity, Weston had to use exposures that were several hours long and so he struggled to produce a clear, still image. Children, household cats and trucks rumbling past the studio all conspired to blur his shell photographs. Over the course of several months, however, Weston eventually produced a dozen or so negatives that he was ready to show people.

Partly inspired by *Nautilus* and several other of his shell photographs, Weston's apprentice, Willard van Dyke, and fellow photographer Ansel Adams in 1932 formed a collective known as Group f/64 (or f.64), which Weston later joined. Their manifesto was to champion pure photography, as they called it, using tiny apertures to produce images of natural forms, landscapes and found objects in a similar style to *Nautilus*.

Weston made 28 prints of his *Nautilus* negative. One was sold in 1927 by a gallery in San Francisco for the sum of $10, on an instalment plan at a rate of 50 cents per month. The photograph was kept in the family until 2010, when the print was put up for auction at Sotheby's in New York. It sold for a little over $1 million.

Back to Wietz's *Shells and Marine Plants*. Among all the large, impressive shells, smaller specimens also catch the eye and have their own stories to tell. There's an aptly named heart cockle (*Corculum cardissa*) to spot. These little white bivalves reveal their heart shape when two shells are articulated together and seen from the side, as in Wietz's painting. In life, these cockles lie in shallow seabeds and through clear windows in their shells they let sunshine stream in, to illuminate tiny symbiotic algae that, much like in giant clams, live inside their bodies and produce sugars to sustain them.

The white open end of a precious wentletrap shell (*Epitonium scalare*) is visible in the cluster of shells in the lower left corner. This shell has unusual and elegant deep ribs which resemble the steps on a spiral staircase (the name 'wentletrap' is borrowed from the Dutch word *wenteltrap*, meaning 'winding staircase'). These tropical shells used to be incredibly rare in the shell-trading marketplace and became legendary among European shell collectors in the eighteenth century. There were so few of them that the rich and famous paid the equivalent of tens of thousands of dollars for single, illustrious specimens. Rumours spread of people crafting counterfeit wentletraps from rice paper, although no existing specimens have been found and this might be a story spread to further inflate the shells' appeal. Until the 1800s, Dutch traders cornered the supply and brought them back to Europe from the Pacific. By the time Wietz painted *Shells and Marine Plants* these shells were already becoming more common. Nowadays, wentletraps are still popular among shell collectors, but there are far more of them on sale and prices are no longer sky-high.

Wietz's shell ensemble also has at least six small cone shells to find. Named for their neat, conical shells, members of the cone snail family (the Conidae) range in size from a fingernail to more than a handspan in length, and they are notorious for being some of the deadliest animals in the ocean. At night-time, these tropical sea snails awake, rising up from their hiding places in the sand,

and begin to roam about hunting for prey. Each of more than seven hundred species of cone snail has its own preference; some are worm eaters; some crab eaters; some eat other molluscs; and some have adopted a most unlikely diet for a slow-crawling snail, and focus on fish, sneaking up on them while they sleep.

The cone snails' weapons are hollowed-out teeth that they fill with complex cocktails of toxins, called conotoxins. They spit out their lethal teeth through a tube, like a poison dart, and following a successful hit the toxins cause instant paralysis by blocking the prey's nerve signals between brain and muscles. Some cone snail species complement this by wafting a cloud of chemicals that mimic the hormone insulin, causing the target's blood pressure to suddenly crash, knocking them out. This gives the snails enough time to crawl up and engulf their motionless prey, swallowing them whole in their immensely extendable mouths. A while later, a fish-eating cone snail will spit out a bundle of undigested bones and scales.

Conotoxins are extraordinarily potent. A couple of cone snail species are deadly enough to bring down an adult human being with a single sting, paralysing the diaphragm so that the unlucky victim can no longer breathe. And being such complex toxins, there are no antidotes. Deployed by the snails, these toxins are fearfully dangerous, but their properties are proving to be immensely useful as the inspiration for new medicines. Medical scientists have extracted and characterized individual conotoxins – there are tens of thousands of these chemicals made by cone snails – and deciphered how they work on the body. One type of conotoxin has been developed into a painkiller that is more effective and less addictive than morphine, and works by blocking nerves that deliver pain signals to the brain. Other medicines in the pipeline based on cone snail toxins include treatments for COVID-19, AIDS, diabetes and malaria.

Alongside their chemical complexity, cone snails also have fabulously elaborate patterns decorating the exterior of their shells, which make them the darlings of shell collectors and mind-bending puzzles for mathematicians and scientists to ponder.

In Wietz's painting, you can make out the intricate triangles and sawtooth marks on the cone shells. It's a little hard to tell, but one of them might be the glory of the sea (*Conus gloriamaris*), with especially fine markings, for a time a rare and highly admired species that, like wentletraps, sent collectors at auction houses into bidding frenzies. A story – again, perhaps apocryphal like the rice-paper wentletraps, but revealing nonetheless – tells of a Danish collector who bought a glory of the sea in 1792 and then immediately smashed the shell to pieces in order to maintain the value of the single specimen he already owned.

Other cone shell species have elegant polka dots, zigzags, bands, stripes and raspberry ripple whorls. Marine biologists have scratched their heads about the purpose of these intricate markings. The snails themselves are unlikely to admire and recognize each other's patterns, because their vision is poor and they're nocturnal, spending daylight hours hidden in the sand and only venturing out under the cover of darkness. It is possible that the shell patterns have no function to speak of, and are merely evolutionary doodles, the result of natural selection running without any constraints on survival.

An alternative explanation proposes that cone snails and other highly decorated molluscs use their patterns as a way to organize their shell-making process. Molluscs are not constantly expanding their shells, but it's an intermittent undertaking, which can speed up when times are good and slow down when food and shell-making materials are scarce. This is why many shells have growth lines etched into them, reflecting the seasonal changes in their shell making. Like tree rings, you can count the furrows on a shell and work out how long a mollusc lived before casting its home on a beach somewhere. Molluscs can likely detect the patterns of pigments in their shells using chemosensory receptors in their soft mantle tissue that can lick over the shell like a tongue. The patterns might remind the mollusc where they left off last time they were augmenting their home, and show them where to line up their mantle and continue secreting calcium carbonate in the correct orientation to neatly expand their spiral. The patterns could be a mollusc's notes to itself – memories, even – jotted down on their shells.

The patterns consist of coloured pigments that are laid down by the mollusc's mantle along the growing edge while new shell material is being made, in much the same way as an inkjet printer sprays drops of ink from nozzles in lines while the paper moves past underneath. Thus the pattern on the shell emerges line by line. The question that has been on the minds of mathematicians is not why cone shells have so many patterns, but how. Patterns are likely controlled by the firing of nerves that can either stimulate or inhibit the secretion of pigments. Computer models have shown that programming fairly simple interactions between these nerve impulses can create complex patterns of stripes, triangles and zigzags spreading across virtual shells. Nobody has yet attempted to study this process in living snails, chiefly because they grow their shells so very slowly.

Below: *Rembrandt,* The Shell, *1650, print.* Opposite: *Wenceslaus Hollar,* Imperial Cone Shell, *1644–52, etching.*

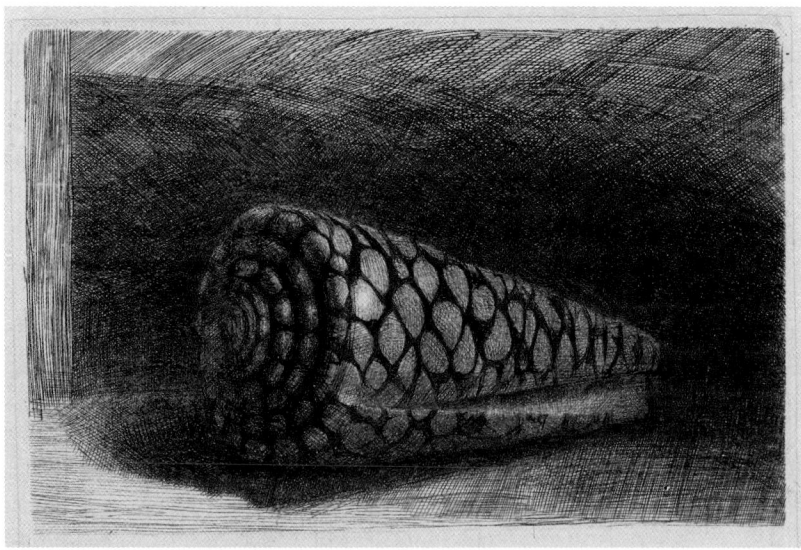

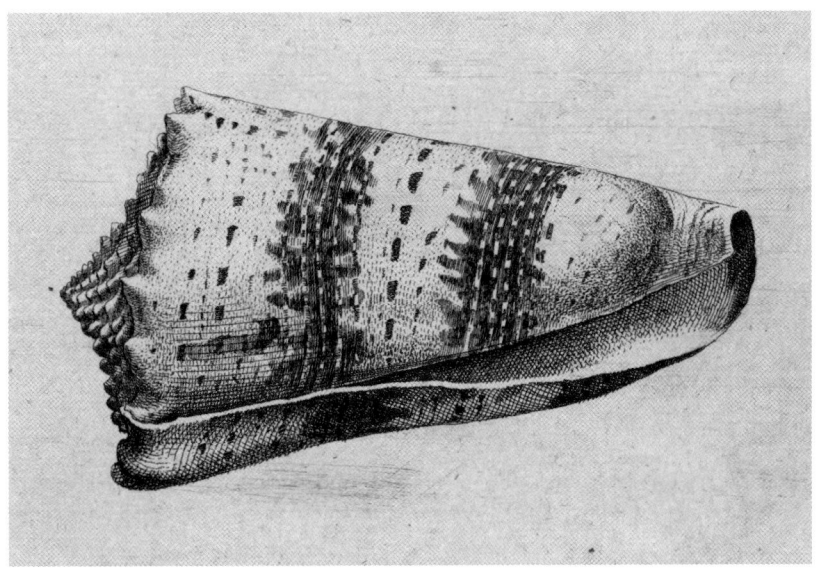

The marbled cone snail (*Conus marmoreus*) from the Indian and Pacific Oceans is another species of celebrated beauty. Their shells have a black or dark red background with large white triangles and tooth shapes across them. Rembrandt's etching of a marbled cone from 1650 is a rare still-life among the Dutch painter, draughtsman and printmaker's works. The sudden appearance of a shell seems to have been sparked by a series of shell etchings made between 1644 and 1652 by Wenceslaus Hollar. Rembrandt likely saw these intricately observed renderings as a challenge to do even better himself.

A marbled cone doesn't appear among the dozens of shells etched by Hollar, although there is an imperial cone (*Conus imperialis*). Hollar and Rembrandt made the same mistake in etching their cone shells. Both shells lie on their sides – Hollar's on a plain white background and Rembrandt's with a darkly shaded surrounding space – and the shells' spires point towards the left, the openings facing upwards. Rembrandt's marbled cone is turned slightly towards the viewer so that we can see the spirals coiling in an anticlockwise or sinistral shell, an orientation which is almost

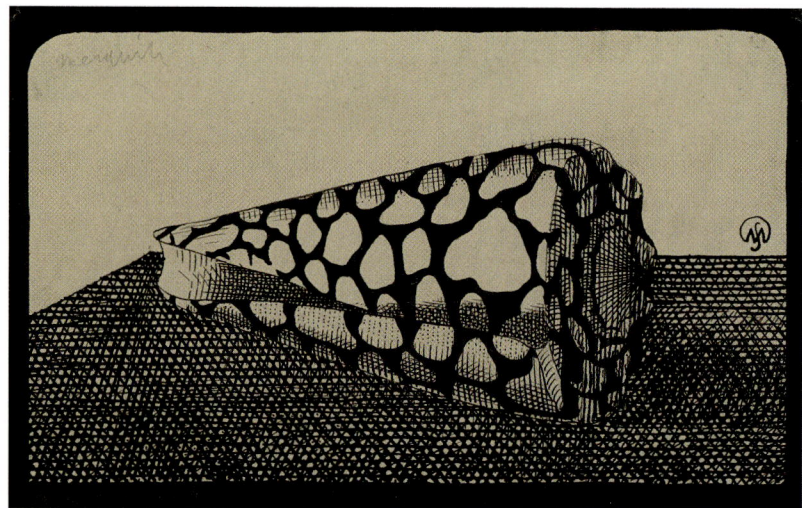

Samuel Jessurun de Mesquita, Cone Shell, Facing Left,
20th century, ink on cardboard.

never seen in nature. All the hundreds of living species of cone snails make dextral or clockwise-coiling shells. The only known exception with an entirely sinistral shell is *Conus adversarius,* a species that went extinct millions of years ago during the Pleistocene.

It is not just cone snails but most types of shell-making mollusc that make dextral shells; find a coiling snail shell, hold it tip-downwards and you'll see that it opens to the right. A curiously similar small proportion of snails as humans are left-handed, roughly 10 per cent. Why exactly right-coiling shells are all the rage among molluscs is not entirely clear, but it likely has something to do with sex. Snails with shells that coil in opposite directions find it physically challenging to mate with each other. Not only are their shells twisted, but their bodily organs are arranged to one side or the other, including their genitals. When snails cling together in a mating embrace, the act is only a success if the relevant parts can come into contact, the male's penis with the female's genital pore (or, for hermaphrodite species, whichever individual is acting in the moment as the male and the female). Like two people reaching to shake hands, or kiss cheeks, it only works smoothly when both offer the same hand or cheek, left or right.

Now and then, due to rare genetic mutations, snails from a dextral species are born with sinistral shells. In 2016, after a left-coiling garden snail (*Cornu aspersum*) was found in London, a national appeal was launched to find them a sinistral mate. Two were located and produced 56 offspring, all with right-coiling shells. The encounter helped scientists identify a gene that produces a sinistral shell in roughly one out of 40,000 garden snails.

As always, the promise of great rarity has made shell collectors eager to get their hands on sinistral specimens. Fraudulent traders have doctored dextral shells, cutting and gluing different parts together to make them look like they coil to the left, then selling them for suitably rare prices.

Rembrandt and Hollar created their sinistral cone shells by drawing on their etching plates precisely what they saw. Then, in the printing process, with paper laid down on and peeled off the inked plate, the etched images were reversed. Of course, artists knew this very well. Rembrandt signed his name and date in mirror writing, so the prints appeared the correct way round. It clearly didn't bother him that the shell was the wrong way round, even though he went to great lengths to observe and accurately depict his life-size depiction of the shell.

The shell that Rembrandt drew would have come from his own valuable shell collection. Six years after he made this etching, the artist went bankrupt, perhaps in part because he spent so much money on shells. All his belongings, including large quantities of shells and corals, were sold at auction. Later on, Rembrandt's etching spawned countless marbled cone shell artworks, made by printmakers and painters who for centuries to come copied the original or made their own versions. One such work, a twentieth-century drawing by the Dutch graphic artist Samuel Jessurun de Mesquita (1878–1944), depicts the shell lying with its spire pointing to right on the page, with its clockwise turning end visible. This is a marbled cone shell as it should be.

Sun-Print Seaweeds

A marbled cone shell appears in another series of shell illustrations, again the right way round, this one dating from 1823. The species was selected to represent the whole cone shell genus (*Conus*). It features alongside a cowry, a bailer shell, a triton shell and many other specimens in an English translation of the French scientist Jean-Baptiste Lamarck's work *Genera of Shells*. The illustrator was a British woman, Anna Atkins, who was then in her early twenties and is now remembered for her groundbreaking works depicting a different group of sea creatures that can be found on the shoreline.

Atkins's seaweed cyanotypes are instantly recognizable. Intricate white shapes look like they've been cast adrift on a flat, sapphire-blue sea. There are lacy, fern-like specimens; flattened blades; mossy clumps; fronds dotted in round berries; and faint, ghostly wisps. All are species of seaweed, also known as algae, that grow around the coasts of Britain.

Today, Atkins's seaweed prints sit somewhere between science and works of art. Originals are owned today by leading scientific institutions and art galleries around the world. During her lifetime, Atkins was very much a scientist, and an unusual one – being a woman. Science was dominated by men, although women were permitted to partake in the genteel discipline of botany. Members of the British public were taking great interest in the natural world, and there was a hazy line between professional and amateur naturalists who were venturing out and finding specimens to study and collect. Many women collected plants and seaweeds and pressed them into scrapbooks. Atkins took this hobby and did something nobody else had done before, pushing it firmly into the world of science.

An only child whose mother died when she was young, Atkins was raised devotedly by her scientist father, John George Children. He was well connected in London's scientific circles and a fellow of the Royal Society, and worked as keeper of the Department of Natural History and Modern Curiosities at the

British Museum. Immersed in her father's world, Atkins stepped into his laboratories for lessons in chemistry and had a thoroughly scientific education. In 1825 she married a wealthy merchant, John Pelly Atkins, and moved into the family home near Sevenoaks in Kent, where she continued to keep the company of various of her father's eminent acquaintances who helped shape her own future. Photographic pioneer William Henry Fox Talbot was a family friend, and in 1841 he gave Atkins a camera, making her possibly the world's first female photographer, although none of her photographs survive.

Her other great influence was the astronomer and polymath John Herschel, who in 1842 invented a camera-free process for creating photographic images, called the cyanotype. The first step involves taking thick sheets of paper, soaking them in a solution of iron salts and leaving them to dry. This makes the paper light-sensitive. Any parts that are subsequently exposed to the ultra-violet rays of the Sun turn a deep, Prussian blue. Any objects laid on the paper that block the sunlight and cast a shadow leave white areas, thus forming a negative white-on-blue image.

Herschel wasn't convinced it would become a useful and popular technique because he wanted to make direct-positive images, ones that looked similar to the original objects or scenes, rather than the reverse as in cyanotypes. The process, however, was adopted by engineers and architects and for more than a century became the main way to reproduce technical drawings. Designs were first drawn onto cartridge paper, then traced with Indian ink onto tracing paper, which was used to make multiple cyanotype copies, otherwise known as blueprints.

Atkins learned and perfected the cyanotype technique and used it to create her greatest works. She took dried and pressed specimens of seaweeds that she had collected herself or borrowed from other collectors, and carefully arranged them on sheets of cyanotype paper. A sheet of glass kept the seaweed flat while she exposed the ensemble to sunlight for thirty or forty minutes. She added titles and labels to each page by placing handwritten notes on slips of oiled paper alongside the seaweeds. Then she rinsed the paper in water, let it dry and saw intricate images appear as if the strands of seaweed were drifting up towards the surface of the ocean. Her pictures were not uniformly blue and white, but many shades of blue, which were produced as sun shone through translucent membranes of the seaweeds, sometimes overlapping in gossamer layers, adding to the aquatic qualities of her images.

In 1843 Atkins self-published her first volume of images. She intended *Photographs of British Algae: Cyanotype Impressions* to

Below: *Anna Atkins*, Asperococcus echinatus, *1850–52, cyanotype print.* Opposite: *Anna Atkins*, Ptilota sericea, *1850–52, cyanotype print.*

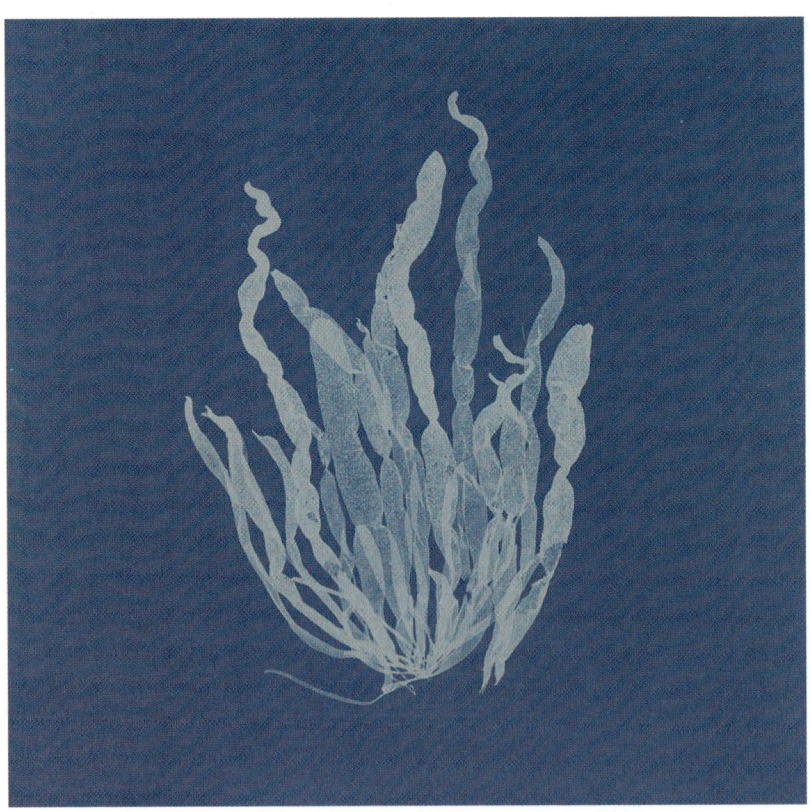

accompany the unillustrated *Manual of British Algae* by William Henry Harvey. Atkins's was the first ever published book of photographs.

Over the following decade, while in her late forties and early fifties, Atkins produced more than twenty volumes filled with thousands of seaweed cyanotypes. She had plenty of material to work from, given that the British Isles are a hotspot of seaweed diversity (to date, phycologists have identified more than 640 species). Each sheet was made individually by her own hand because there was no way to automate or duplicate the process. No two images are exactly the same. She gave collections to friends and

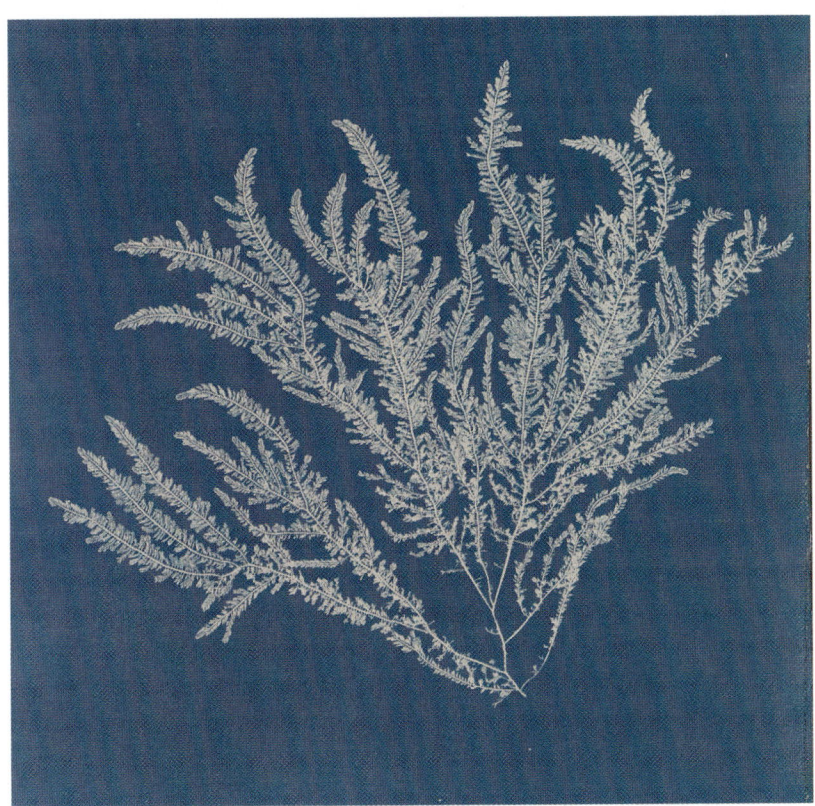

museums, the pages stitched loosely together by hand and never bound firmly into hardback volumes. The copy now kept at the New York Public Library had been a gift to Herschel.

Atkins went on to produce cyanotype works featuring ferns and other land plants, but today she's most celebrated for her seaweeds. Cyanotypes didn't take off as a way to illustrate scientific texts, probably because they lack important details such as the colour and surface texture of specimens. And yet, until advances in photography in the twentieth century, Atkins's approach was the only method of capturing an accurate image of a wild species that did not depend entirely on an illustrator's eye. Hers was a revolutionary way of knowing the world and it diverged from the engrained view that mankind is set apart from everything else and acts as a dominant force observing, cataloguing and controlling the natural world. Instead, Atkins worked alongside nature and, with just a little careful arrangement, she let natural objects make their own impressions.

Beached Behemoths

Far more ominous than the seashells and seaweeds that wash up on coastlines are the ocean's largest animals – and the biggest creatures ever known to have existed – that sometimes find themselves stranded on beaches, collapsing under gravity and surrounded by onlookers. A record of one such animal remained out of sight for more than century. *View of Scheveningen Sands*, painted around 1640 by the Dutch artist Hendrick van Anthonissen, depicts an unassuming scene of a flat sandy beach and sand dunes beneath a cloudy sky, a few small boats pulled up onto shore, cart tracks running along the beach and a few dozen people standing around near the gently breaking waves, for no apparent reason. This was how the painting looked in the 1870s when it was acquired by the Fitzwilliam Museum in Cambridge, England. In 2014 art conservationist Shan Kuang was restoring the painting, removing the yellowed varnish from its surface, when she began to notice the faint silhouette of a human figure that seemed to be standing on

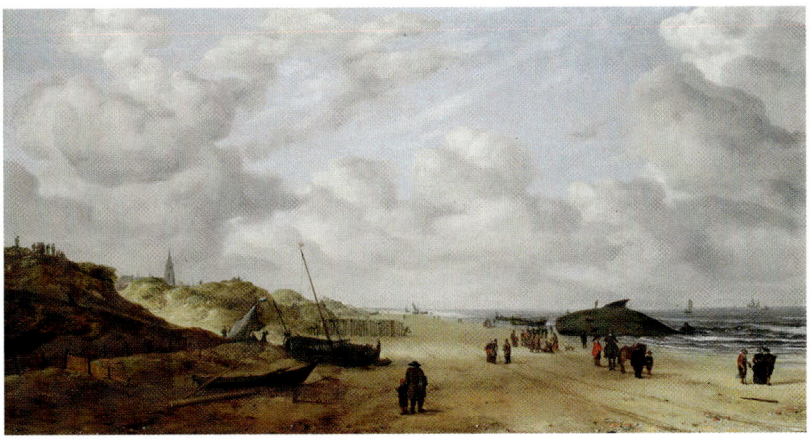

Hendrick van Anthonissen, View of Scheveningen Sands,
c. 1640, oil on panel.

the sea, just above the horizon line. She realized that the painting
had been altered and set about using a scalpel, solvents and a
microscope to carefully scrape off the overpaint. Underneath, she
found the hidden body of a stranded sperm whale (*Physeter
macrocephalus*) lying with its head towards the beach, and a man
balanced on its back holding a line presumably measuring the animal's immense length. This was why all these people had gathered
on the beach, to witness a spectacle from the ocean.

Manipulating artworks was an acceptable practice in earlier
centuries when art dealers had no qualms in cutting out parts of
paintings or having more paint added in order to create images
that fitted with tastes of the time and were more appealing to
buyers. Experts haven't been able to work out who painted over
Anthonissen's whale, or when exactly they did so. But it was obviously from a time when dead whales were no longer the artistic
draw they once had been.

In the Netherlands in the early seventeenth century there
was a frenzy of public interest in dead whales washing up on
beaches, a natural occurrence that happened from time to time.
People flocked to see these immense beasts that appeared from the

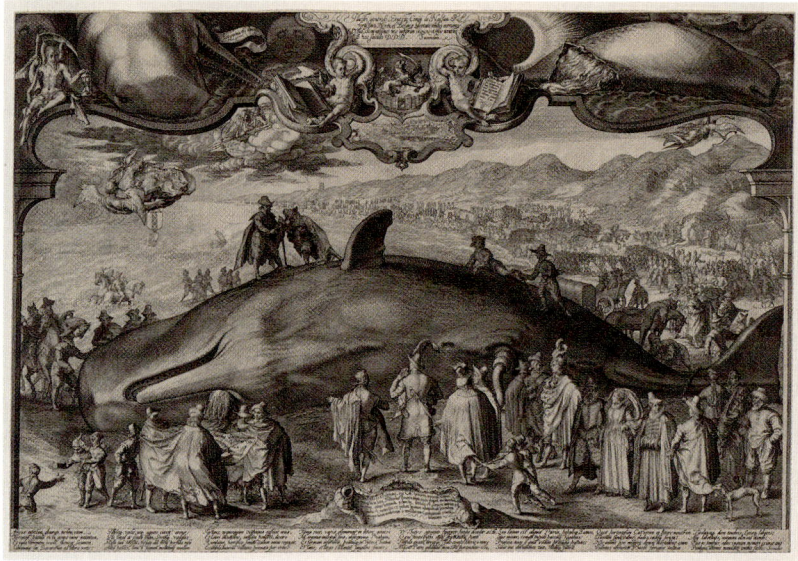

dangerous and mysterious ocean depths. Artists depicted these scenes in numerous paintings and engravings, no doubt urged by the popular belief that beached whales were bad omens and warnings of troubles to come.

A 1602 engraving by Jan Saenredam shows a stranded sperm whale that appeared on a beach near Beverwijk on the coast of North Holland in December 1601. The whale lies on its side with its belly towards the viewer. The narrow jaw hangs ajar, showing a neat line of dark dimples (people used to think these were the holes where a sperm whale's upper teeth had fallen out, but in fact this whale species only has teeth in its lower jaw; these whales suck up their prey, mostly soft squid, without chewing). Hanging from the whale's mouth is what appears to be a tangled part of the regurgitated stomach. Further along the body is a part of the sperm whale's anatomy that is usually not on display in living animals but kept tucked inside until it's needed, keeping the body streamlined. The Beverwijk stranding was evidently a male whale; as the carcass began to decompose and release gases, the pressure inside increased and pushed out the whale's large penis so that it dangled down onto the sand.

The whale is surrounded by crowds of well-dressed noblemen and -women, and a mass of armed soldiers, their dogs and horses all standing along the beach, which stretches into the distance. As in many similar scenes, people have climbed on top of the whale, perhaps to get a good view and wave at their friends. A pair of men are stabbing a sword into the whale's blowhole. A rope is slung around the base of the whale's tail, indicating that it may have been hauled up the beach, perhaps as the tide was rising.

Two identifiable individuals stand in the foreground. One is the artist himself, Saenredam, sketching on paper laid on a wooden barrel and using his cape as a windbreak. To Saenredam's right stands Count Ernst Casimir of Nassau-Dietz sporting a plumed hat and delicately holding a handkerchief to his nose in an attempt to overcome the terrible stink of the rotting whale. It's possible the count had daubed the handkerchief in a substance that had likely come from a similar animal.

Opposite: *Jan (Pietersz.) Saenredam,* Beached Whale at Wijk aan Zee, *1602, engraving.* Below: *Etching of ambergris, from Georg Eberhard Rumpf,* D'Amboinsche Rariteitkamer . . . *(1705).*

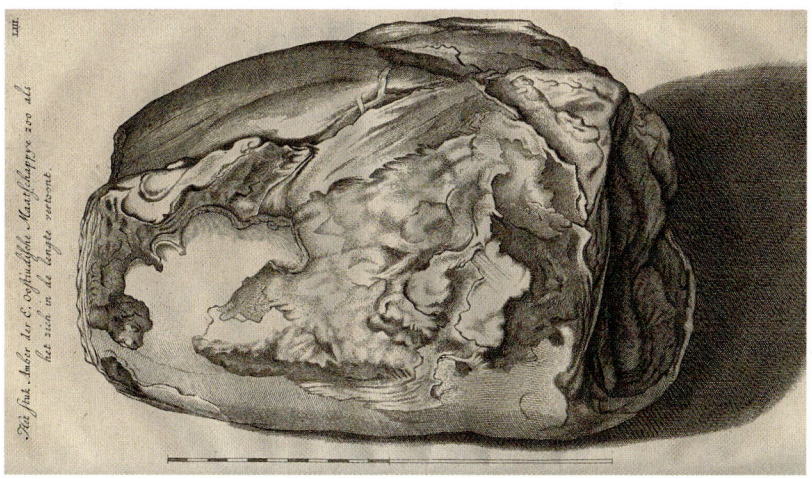

Ambergris is a rare and treasured substance that people have known of and used for centuries for its ability to fix aromas in perfumes and for its own unique earthy, woody scent. In early modern Europe, it was generally thought that diseases were transmitted through the air as bad smells, and could be blocked by more pleasant smells. Ambergris was burnt as incense to fumigate homes and placed inside pomanders with aromatic spices to protect against the plague.

The origins of ambergris were a long-standing mystery. Did it come from underwater mountains, or was it hardened chunks of sea foam or seabird droppings? Only when commercial whale hunting gathered pace in the 1800s and thousands of animals were killed did it become clear that ambergris originated deep in the guts of sperm whales. Fewer than one in ten sperm whales have a constriction in their lower bowels that makes it difficult for them to pass the undigested remains of their meals. Hard beaks of squid can form a congealed nugget, smeared in layers of oily secretions, that is eventually released into the sea either when the whale manages to push it out in its faeces or when the animal dies. The floating mass then gets chemically altered by the salt and sun, converting the faecal aromas into something altogether more complex and appealing. Today there is a synthetic version of the key ingredient, a molecule called ambrein, which is widely used by perfumers. The trade in wild ambergris is now widely restricted, due to the protections on great whales and anything that comes from them. But still people look out for valuable lumps of ambergris that infrequently wash up on beaches.

Surrounding Saenredam's stranded-whale scene, an illustrated frame refers to various forbidding events that unfolded in subsequent years and which the ill-fated beast had surely foretold: the city of Amsterdam was struck by the plague, there was an eclipse of the Moon and of the Sun, and a great earthquake struck off the Netherlands coast.

Beliefs have persisted into more recent time that earthquakes are forecast by certain oceanic animals stranding on beaches, such as long, silvery oarfish (from the family Regalecidae).

Scientists have investigated this idea and searched for records of oarfish strandings but found no overlap with the timing of earthquakes. Out of more than one hundred confirmed reports of oarfish washing up, the nearest any came was one fish that showed up within thirty days and 100 kilometres (62 mi.) of an earthquake's epicentre, putting to rest the idea that these great long fish sense underwater seismic waves and respond by rushing to the sea surface and casting themselves ashore.

What causes animals to strand on beaches remains quite mysterious and certainly complex, especially the tragic mass strandings of cetaceans such as pilot whales that sometimes charge onto the shore in their hundreds. In some cases, a sick or injured individual gets in trouble and then these highly social animals cluster near to help but end up all getting caught in shallow waters and stranded together. Booming underwater sonar used by the military in anti-submarine training exercises can lead to whale strandings, as can oil and gas corporations in their search for new seabed deposits. The deafening sounds make whales panic and swim to the surface too fast, so bubbles of gas form in their blood, like opening a shaken can of beer. The same debilitating and often lethal condition, called decompression sickness or the bends, happens in human scuba divers who surface too quickly from a long, deep dive.

And while earthquakes and solar eclipses are probably not involved, a plausible link does indeed exist between other astronomical events and whale strandings. Around Christmas 2015, the skies of northern Europe lit up with spectacular displays of the aurora borealis. At around the same time, the biggest mass stranding of sperm whales on record took place. Within a matter of weeks, 29 whales had washed up dead on beaches in the Netherlands, Britain, Germany and France. They were all young male whales and all in good health.

The North Sea is one of a handful of places in the ocean that commonly traps whales, likely because the water body is enclosed and relatively shallow, getting progressively shallower towards the south. This seems to confuse the giant cetaceans and is likely why so many have stranded here, including on Dutch shores centuries ago. The Christmas strandings of 2015 could have been triggered by the same phenomenon that illuminated the skies. Coronal mass ejections on the surface of the Sun were hurling charged particles at Earth and stirring the northern lights. The same solar storms substantially shifted Earth's magnetic field, which may have confused migrating sperm whales.

It has yet to be proven, but it is possible that whales join the likes of sea turtles, bees and salmon and can sense magnetic fields and learn to use the slight differences across the planet's surface as maps to navigate from place to place. When solar storms push these magnetic maps out of alignment, migrating animals may unwittingly veer off course. Studies of whale strandings in the North Sea have found that they occur more frequently in years with more solar activity and frequent solar storms. One theory about the 2015 strandings is that the young sperm whales that would normally have swum north into their hunting grounds of the Norwegian Sea instead were tricked by the misaligned magnetic fields and swung south, ending up in the whale trap of the North Sea. And so the wintertime skies glowed in luminous greens and blues that reflected off the bodies of so many sperm whales that fatally left their underwater realm.

Beachcombing in the Anthropocene

In the early decades of the twenty-first century, with the human-dominated Anthropocene well under way, beachcombers are no longer hunting just for nature's treasures and the captivating remains of sea creatures cast up on the shore. Now, woven among the seashells and seaweeds are pieces of colourful, durable trash that people have discarded into the ocean – deliberately or un-thinkingly – and have floated back towards the land masses where they were made. For many, the pastime of beachcombing has become an endeavour of beach cleaning.

Plastics manufacturing has been exponentially growing since the mid-1900s. In recent decades it has skyrocketed largely

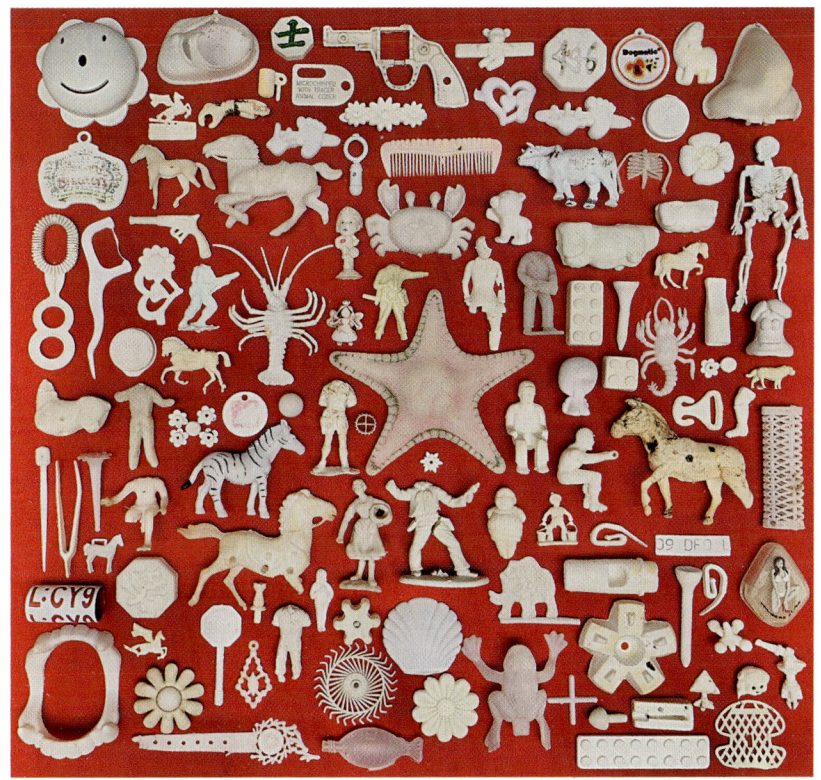

Tracey Williams, tableau of beach-combed plastics, 2020.

because fossil-fuel companies have decided to turn more of their oil, gas and coal feedstocks into plastic bottles, packaging and other items that are used just once and then thrown away. Presently, roughly 400 million tonnes of new plastics are made every year, and less than 10 per cent of them are recycled. A lot ends up in the ocean.

Shocking images of plastic junk choking and killing ocean wildlife have become a core part of the growing culture of environmental awakening. Footage of a sea turtle with a drinking straw pushed up its nose, a seahorse drifting through the sea with its tail wrapped around the plastic shaft of a cotton ear swab and dead seabirds with their insides stuffed with plastic scraps and cigarette lighters have helped to make people take notice and care about the issue of plastic pollution.

This ocean problem has also spawned imagery that holds a compelling, if disturbing, beauty. Some beachcombers have taken to photographing tableaus of the plastic objects they collect arranged by shape, size, type and colour. These orderly ephemera are at first glance pleasing to the eye, but become more unsettling the longer you contemplate what they are and where they came from.

In the late 1990s, beachcombers searching the flotsam and jetsam on beaches of southwest England began to spot thousands of pieces of the classic Danish children's plastic building blocks Lego. Curiously, a lot of the beach-stranded Lego pieces were sea-themed. There were tiny plastic life jackets, scuba tanks, ship's rigging and fronds of seagrass, plus the occasional octopus and shark.

The source of the Lego has been traced to a cargo ship, the *Tokio Express*, which in February 1997 got into trouble in a violent storm more than 30 kilometres (20 mi.) off the western tip of Cornwall, which pokes out into the Atlantic. Enormous waves swept 62 metal containers off the ship and into the sea. Inside one of them were close to 5 million pieces of Lego, which were on their way to the United States to be put in seafaring toy sets. That single container spilt its contents and created a slick of Lego pieces, many of which floated towards the Cornish coast.

Among the beachcombers gleaning aquatic Lego is Tracey Williams, who early on decided to post pictures on the Internet of findings from her local West Country beaches. She soon gathered an expanding following on social media, including oceanographers, archaeologists and other plastic-pickers around the world who help identify her pieces and share their findings of Lego that are showing up further afield. Some types of Lego figurines wash up frequently because tens or even hundreds of thousands of them were inside that spilled container. Flippers for Lego diver figurines are especially common, meanwhile black octopuses are rare finds.

More than twenty years later, the *Tokio Express* Lego pieces are still washing up; some must have got caught in swirling currents and only just reached shore, and some likely sank and were swept on deeper undersea currents near the seabed. In 2022 Williams published *Adrift: The Curious Tail of the Lego Lost at Sea*, a book of photographs which many readers appreciate as intriguing works of modern art. Her social media feeds show her ever-expanding collections of Lego and other plastic beach finds. There are Lego dragons and broomsticks, My Little Ponies, Peppa Pigs, toy soldiers, fake flowers, toothpicks, nose sprays, circular wands for blowing soap bubbles, and those takeaway sachets for soy sauce that are shaped like fish. Williams has said that she doesn't consider herself to be an artist and insists that her works are not intended to be beautiful. Rather, she regards all these found objects as ingredients in a dangerous and thickening plastic soup in the ocean. She wants her images to help make tangible the links between human lives on land and the junk that one way or another ends up in the ocean.

TWO

SHALLOWS

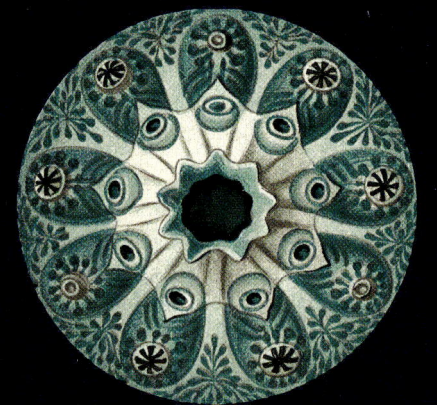

Stepping across the tideline, we wade into the shallow seas that fringe the land and lie above the continental shelves. These shelf seas vary in width and also depth, although rarely they exceed 100 metres (330 ft) before reaching the giant cliffs of the continental slope that plunge into greater depths. These are the most accessible waters of the ocean, the places that humanity has interacted with for the longest time, fishing from shore and setting off in sailboats on coastal journeys. Unsurprisingly, the animal inhabitants of these seas that appear most commonly in works of art and hold the most significant meanings for people are fish.

In every coastal culture, and many far inland, fish are food. They have been such important sources of sustenance that people have long considered fish to be symbols of abundance and well-being, and messengers from the ocean's life-giving waters.

There are tens of thousands of species of fish alive today, a roughly equal number living in fresh water as in the salty seas (and many that migrate between the two), and they come in all manner of shapes and sizes: serpentine eels and spherical pufferfish, wafer-thin flatfish and cubic boxfish. And yet the generic shape of a fish holds a pleasing simplicity that is unmistakable and easy to depict: an oval body and at one end a triangular or a fork-shaped tail. All that's needed to draw the discernible shape of a fish, as in the Christian *ichthys* symbol, is two intersecting arcs. Everyone knows a fish when they see one.

Even modest representations of fish can be recognizable as particular species. One of the most ancient known rock art fish was found in the late nineteenth century in the Vézère valley of southwestern France, in a small prehistoric cave known as the Abri du Poisson (the 'fish shelter'), located 24 kilometres (15 mi.) from the famous cave paintings of Lascaux. Dated to at least 23,000 years ago, this is a rare depiction of a fish in rock art from the last Ice Age, a time when humans were known to be catching and eating them. At a metre long, this is a life-sized carving of a salmon that had migrated from the ocean to their spawning grounds far upriver.

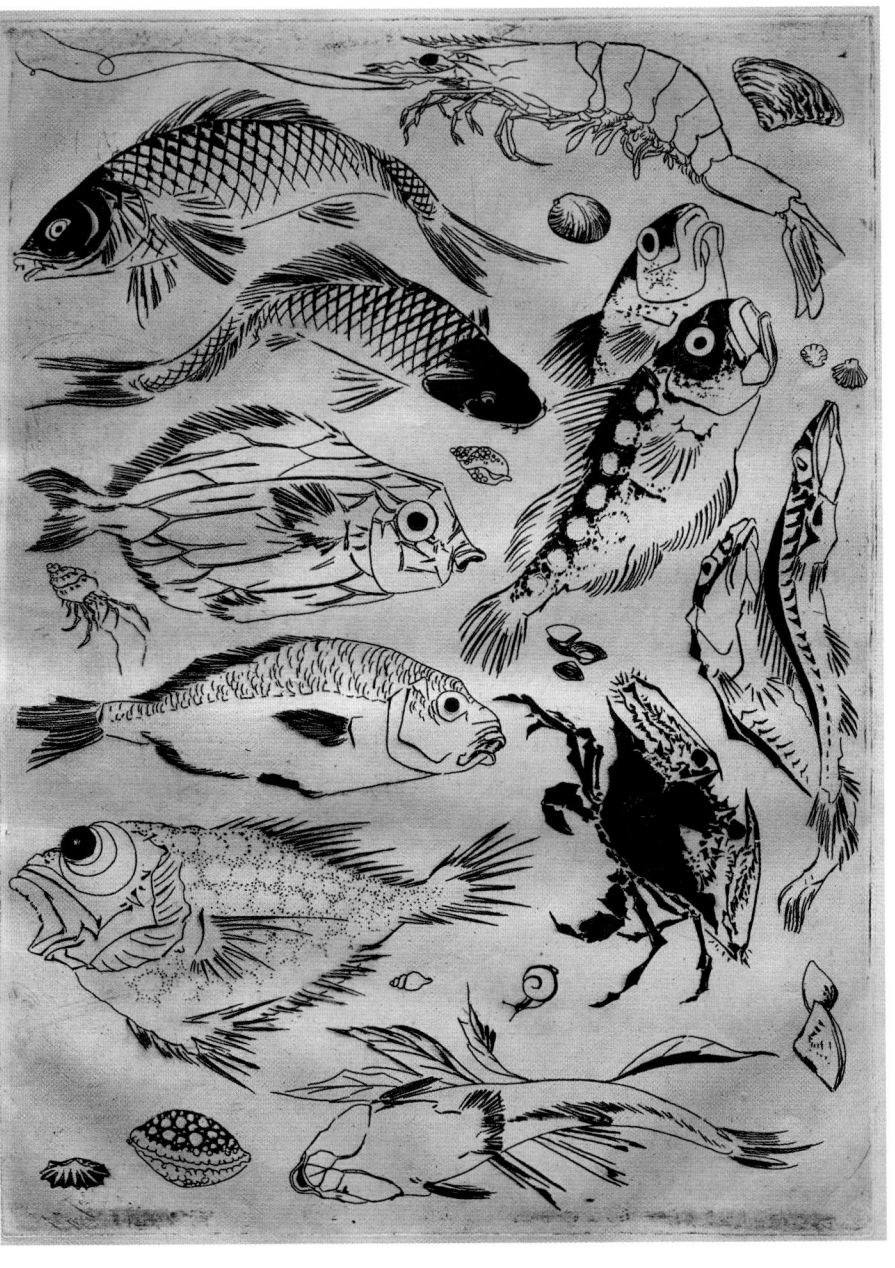

A piece of fish art was created at a similar time in La Pileta cave in Andalusia, southern Spain, not far from the Mediterranean coast. In the deepest part of the cave system is a beautifully observed sketch of a flatfish, likely a halibut. The image has an asymmetrical head with what seem to be two smudged eyes looking out from the rock face; a pair of fins are correctly placed on the body for a halibut. At 1.5 metres (5 ft) from head to tail, the drawing is perhaps smaller than the real-life animal, which could have been more than 4 metres (13 ft) long. How the Palaeolithic artist knew about halibut is rather a mystery, because the Atlantic species generally only lives in water more than 50 metres (164 ft) deep, although they do occasionally rise into shallower seas.

There is something endearingly ichthyological about the stone palettes commonly found in tombs from the predynastic period in ancient Egypt, around the fourth millennium BC.

Opposite: *Félix Bracquemond,* Dinner Service (Rousseau Service): Fish (No. 24), *1866, etching.* Below: *Life-size carving of a salmon, cave art from Abri du Poisson (Fish Rock Shelter).*

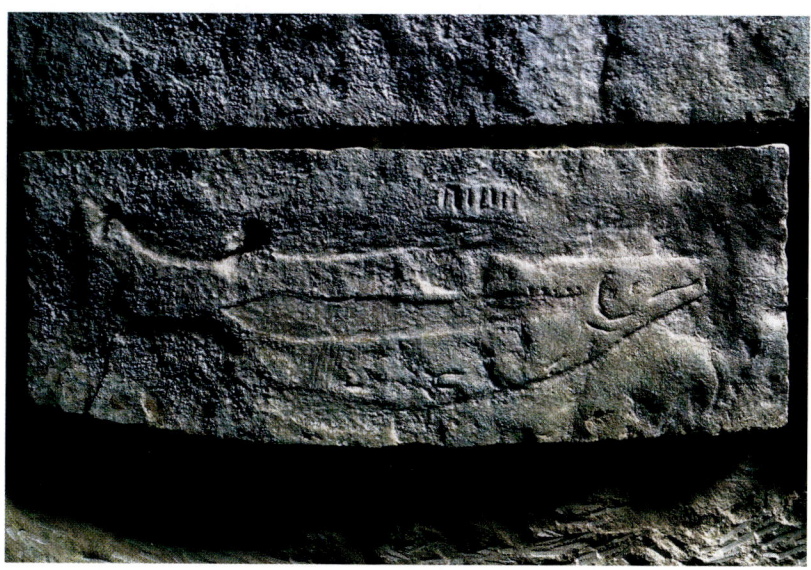

Cosmetic palettes were used to grind black and green pigments into powders to use as make-up. Many were made specifically for burials, as grave goods, and left in tombs for the dead to use in the afterlife. Artisans who carved the palettes captured a fishy essence in the rounded bodies, curved gill cover (or operculum) on the side of the head, simple tail and carefully drilled dot for an eye. The fish represented Nile tilapia, a freshwater species from rivers and marshes, which the ancient Egyptians associated with fertility, growth and rebirth, in part because the female tilapia brood unhatched eggs in their mouths then release a shoal of baby fish, as if they were breathing new life into the water.

A small fragment from a rock crystal bowl found in Tunisia, North Africa, is carved with numerous charismatic grumpy-looking fish with downturned mouths, triangular pectoral fins and fan-shaped tails. Dating from between the third and fifth centuries, the intact bowl may have been suspended by chains, filled with oil and used as a lamp, light flickering through the transparent fish.

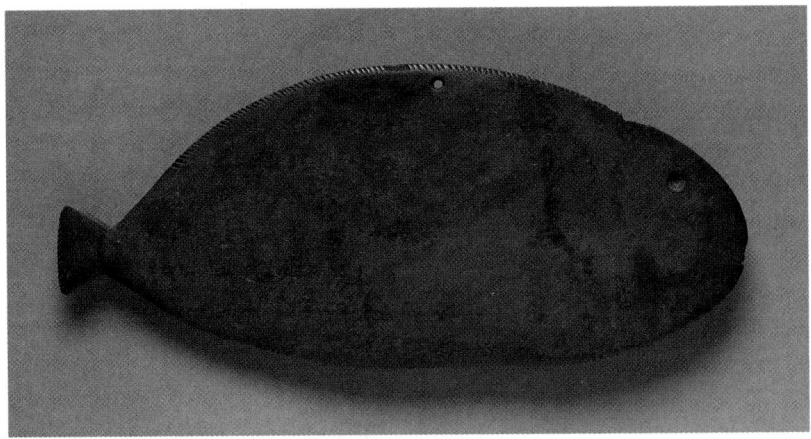

Above: *Palette in the form of a fish, Egypt, c. 3500–2950 BC, sandstone.*
Opposite above: *Fragment of a rock crystal bowl, North African (Carthage), 3rd–5th century.* Opposite below: *Set of three fish, House of Fabergé, c. 1890, agate.*

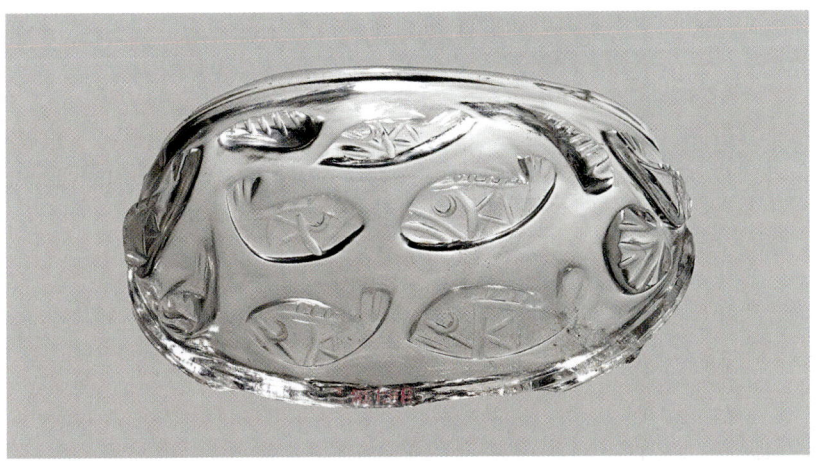

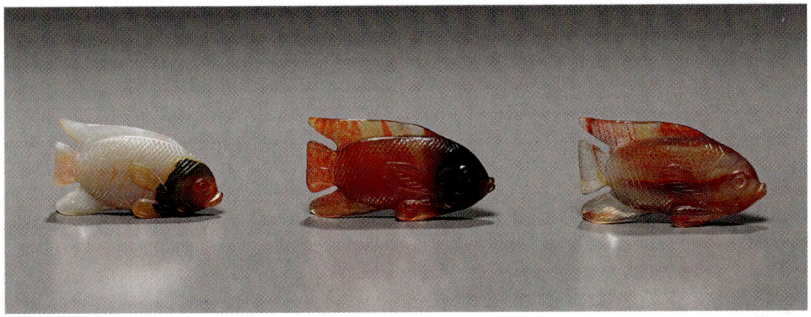

Jumping ahead many centuries towards the present day, fish continue to appear in contemporary artworks, shown in outline as non-specific symbols of aquatic life and not intending to be any species in particular. Among them, I have a special fondness for *Fish Magic* by Paul Klee. The Swiss-born artist created this work in 1925 while teaching at the German art school the Bauhaus, and a while after he visited Tunisia and came back saying he was now possessed by colour. For *Fish Magic* he laid down multiple layers of coloured paint, covered them over in black, then scratched through to reveal the kaleidoscope underneath. Different colours appeared depending on how deep he went, which feels to me like a fitting analogy for exploring the ocean – as you go deeper, the colours change and there are different types of creature waiting to be seen. *Fish Magic* features a mysterious mix of aquatic creatures, human caricatures and flowers. The fish are simple oval shapes with triangle tails, but to me they are nonetheless full of charm, the kinds of creature that I imagine catching my attention underwater before slipping away out of sight.

Sharks feature in artworks and objects around the world and are clearly distinguishable from other types of fish. There are around five hundred living species of shark and they're generally conservative in their body plans. Their typical profile comprises low-slung jaws bristling with triangular teeth, rows of gill slits on either side of the head (there is no gill cover), crescent-shaped tails and multiple pairs of fins arranged along their torpedo-shaped bodies.

Below: *Paul Klee,* Fish Magic, *1925, oil and watercolour on canvas.*
Opposite: *Shark pendant, Chiriqui culture, 11th–16th century, gold.*

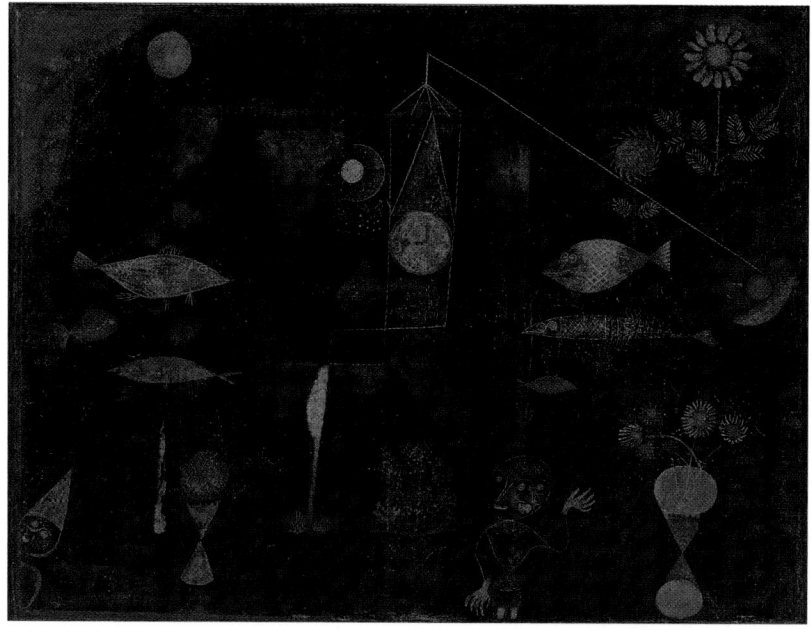

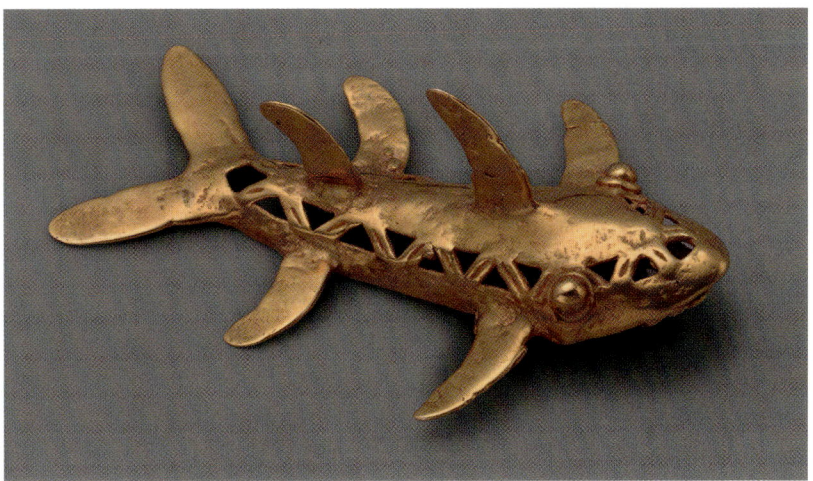

Even in abstract form, these shark characters are discernible, as in the painted ceramics from the Nazca culture, which flourished on the southern coast of Peru from the first century BC into the first millennium AD. A ceramic sculpture from the Tumaco-La Tolita culture of Colombia and Ecuador (first to fifth centuries) has the upturned, pointed snout and fearsome toothy grin of a shark lunging in for a bite. A small shark with its tail curved towards its head, perhaps one of the coastal species that rests on the seabed, was pressed from a thin sheet of silver and used as a textile ornament by people in the pre-Incan Chimú culture of Peru. Not a great deal is known about the Veraguas or Chiriqui culture that existed in Panama and Costa Rica from the eleventh to the sixteenth century, but the people left behind intricate gold ornaments, many cast into the form of animals, including beautiful shark pendants.

Books of Fish

Beyond the decorative motifs of fish and their use as symbols of abundance, wildness and spirituality, there are centuries-old images of fish that trace changing ideas of what kind of animals

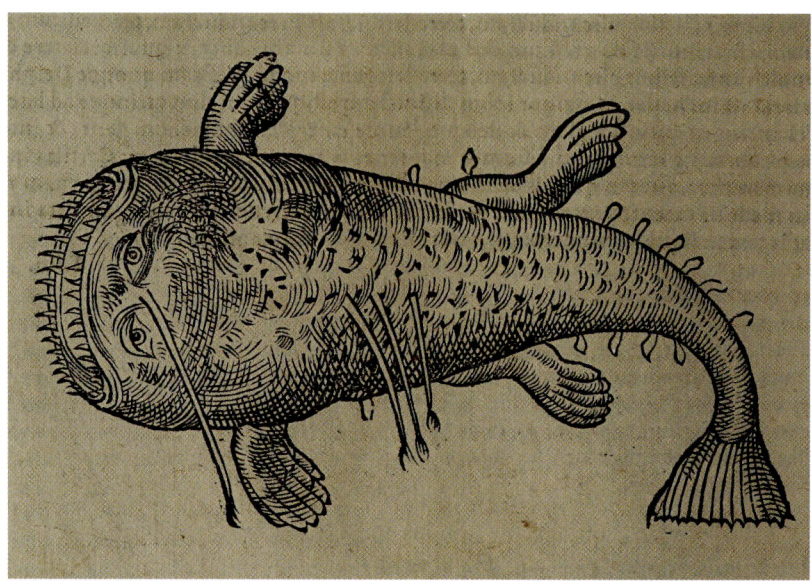

Above: *Monkfish, woodcut from Pierre Belon,* De aquatilibus *(1553).*
Opposite: *Flying gurnard (*De Hirundine*), woodcut from Guillaume Rondelet,* Libri de piscibus marinis *(1554).*

they are, and show a gradual shift towards realistic portrayals of them.

In ancient Greece, the philosopher Aristotle carefully studied many aquatic animals and came up with some thoroughly modern, scientific ideas for how to classify them. He defined fish as blooded animals, as distinct from bloodless organisms like jellyfish (the vertebrates and invertebrates respectively). Specifically, fish are animals that live in water; they have no legs, hair or feathers; their bodies are usually covered in scales, although not always; and they either lay eggs or give birth to live young. Aristotle also understood that fish are different to the kind of animals that he called cetaceans – the whales, dolphins and porpoises – which share similar features with land-dwelling mammals, such as having lungs to breathe air and producing milk to suckle their young.

Much of Aristotle's logical classification became muddled and forgotten over the centuries. Images of monstrous, fish-like creatures appear in the margins of medieval manuscripts and maps, and illustrated books of beasts. Animals that are otherwise recognizable as fish were adorned with human arms and legs, angelic wings and unicorn horns. Hybrids of fish and parts of other animals, humans included, were also popular, perpetuating an idea going back to the first century and the Roman writer Pliny the Elder that every land animal has an equivalent in the ocean: hence there were sea pigs, sea dogs, sea snails, sea horses and so on.

This mingling of myths and realities can be seen in the oldest surviving European books devoted to the study and identification of fish, which started to return to a more scientific way of thinking. In the 1550s, several illustrated fish volumes were published by writers who were all scholars of the living world and who knew each other and likely shared their ideas about the aquatic realm. One was Frenchman Pierre Belon, who produced *L'histoire naturelle des estranges poissons marins* (The Natural History of Strange Sea Fish) in 1551 and *De aquatibilis* (On Aquatic Animals) in 1553. These books included descriptions and

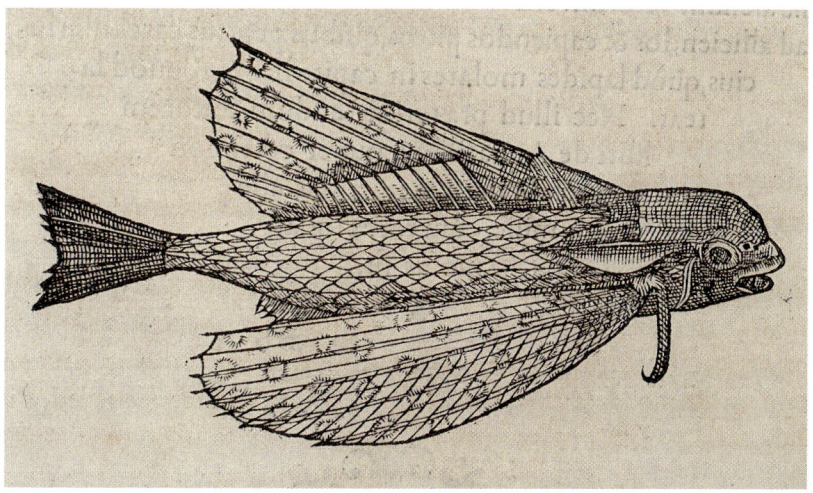

woodblock illustrations of realistic, recognizable species, such as sturgeon, hammerhead sharks and flying fish. Belon also kept in his books some of the cetaceans and other air-breathing animals like crocodiles and hippos, as well as several sea monsters from medieval fables and bestiaries, such as the fish-headed bishop. In 1554, Belon's fellow countryman Guillaume Rondelet published *Libri de piscibus marinis* (Book of Sea Fish), which had more than two hundred species and a similar mix of the real and the imaginary; there are triggerfish, sunfish and flatfish, and a scaly lion with a human face.

Belon's and Rondelet's woodblock fish are undoubtedly charming and many accurately depict attributes that would help readers identify species for themselves, making these titles an early form of natural-history guidebook. Nonetheless they remain caricatures of wild species, which is not the case for another book of fish published in 1554. Italian scholar Ippolito Salviani produced *Aquatilium animalium historiae* (The Histories of Aquatic Animals), which contains astonishingly lifelike copperplate engravings of 96 types of fish, and a distinct lack of mammals and mythical sea monsters. The high production values of Salviani's book allowed him to show what these animals looked like in exquisite detail. Many of the specimens could be alive and ready to slither off the page at any moment. There are moray eels and stingrays, mullet, wrasse and rockfish.

Just over a century later, some of Salviani's fish appeared again in another illustrated book – this one published in London – which earned itself a rather infamous reputation at the time. It began as a joint project by the British naturalists Francis Willughby and John Ray, who spent several years of the 1660s travelling around Europe searching for species of many kinds – birds, insects and fish – with the aim of producing a pioneering series of books about animal life. They gathered specimens and bought books, met local experts and carefully observed species in the wild. Shortly after returning to Britain, Willughby died of pleurisy, leaving Ray to complete their work. He first wrote a book about birds, then moved on to the fish.

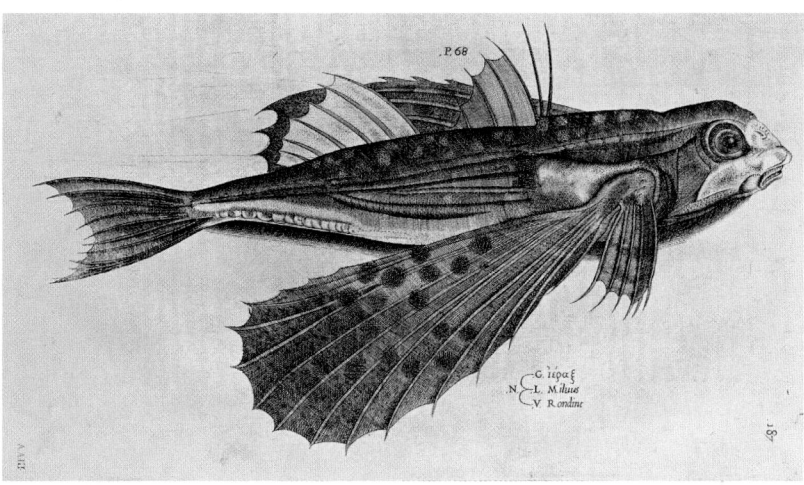

Flying gurnard, woodcut from Ippolito Salviani, Aquatilium animalium historiae *(1554).*

Ray was highly selective about which animals made it into his fish book. He had no time for fish bishops or mermaids, and stuck closely to a more Aristotelian view of fish life. He was also determined to avoid duplications, as had become a habit of many naturalists at the time who were keen to split animals into as many species as they could muster. Ray lumped these synonyms back together and eventually presented 420 fish species in what amounted to the most accurate guide to European fish of its time, *De historia piscium* (On the History of Fishes), published in 1686.

Originally, Ray had not planned for the book to be illustrated. London's leading scientific institution, the Royal Society, had pledged to publish the tome, and fellows at the society helped Ray to revise and correct the manuscript. They persuaded him that it would help readers identify fish species if they had pictures to look at. Illustrations were commissioned, although not from scratch, but, as was commonly done, they were to be copied from existing works.

In total, there were 187 fish depicted in beautiful, large-format plates, some of them evidently copied from Salviani. These were expensive to produce and funded largely by Royal Society fellows who sponsored individual fish. Plates are inscribed with the name of the sponsor of each species, including such illustrious people as Samuel Pepys, Edmond Halley and Christopher Wren. They all paid £1 per fish (equivalent to roughly £200 today), in a seventeenth-century predecessor of crowdfunding initiatives. The fellows' sponsorship wasn't enough to reach the target sum, however, and the Royal Society ended up paying the rest of the bill to produce five hundred copies of Ray's book.

As it turned out, there just wasn't enough demand for beautiful fish guidebooks. The Royal Society didn't sell enough copies to make their money back and the institution almost went bankrupt. It meant that the society had to renege on its commitment to publish one of history's most important science books, Isaac Newton's *Principia*. His book was eventually published, funded privately by Edmond Halley, and it went on to revolutionize the understanding of the universe, although for a while the *Principia*'s future had been threatened by a printed shoal of very expensive fish.

Dazzling Displays

While the science of ichthyology was gaining momentum, aided by carefully observed illustrations, another artist was focusing less on biological accuracy and more on the commercial possibilities of painting charming pictures of fish. At the turn of the eighteenth century, Samuel Fallours was a British-born soldier in the Dutch East India Company stationed on what is now the Indonesian island of Ambon. He was also a fine painter and became enthralled by the colourful fish living in the tropical waters around the island. Local fishers brought him specimens to draw from and he produced paintings to sell to wealthy collectors back in Europe.

Fallours's fish were the basis for a book published in the Netherlands in 1719. Differing from previous books of fish, which featured species from temperate seas and the warm Mediterranean, for European readers these were unfamiliar fish from tropical waters. Only a hundred copies of *Poissons, écrevisses et crabes* (Fish, Crayfish and Crabs) were printed. In the rare copies still known to exist today, there is a miscellany of hundreds of sensational, kaleidoscopic animals. At first glance, these animals are bursting with charisma and look more like cartoons than real fish. And yet, among grinning faces painted with elaborate stars, zigzags and spots there are signs that these were originally intended to be real species. There's a painting of a majestic masked angelfish (*Pomacanthus navarchus*) with an elegant stripe beneath its eye, a trigger fish (Balistidae) with a distinctive prong sticking up from its back and a lionfish (*Pterois miles*) with long, poison-tipped spines.

Although they're not exactly correct, the colours and patterns of Fallours's fish are not so wildly exaggerated. Genuine fish species from the coral reefs of the Indian Ocean and the Pacific are some of the most brazenly colourful animals in the sea. Their markings serve various purposes. A dark mask drawn across the eye helps to disguise this critical part of the body, often paired with a bright, decoy eyespot near the tail, which deflects the attention of predators and confuses them when the fish swims off in apparently the wrong direction. Another possibility is that the bright colours evolved as a way for fish to show off and attract mates. Colour combinations commonly seen on fish are highly visible in the light conditions on shallow tropical reefs. Golden yellow and indigo blue lie on opposite sides of the artist's colour wheel, and are seen in the bold colours of angelfish and butterflyfish, making them stand out when seen from a distance on a reef.

Most of all, striking colours declare a fish's identity, with the intended audience being other members of the same species, which on coral reefs are often highly territorial. Scientists refer to them as poster colours. The idea is that these bold, striking motifs are an obvious sign to potential competitors that this part of the reef has already been claimed. This theory is strengthened by the fact that many coral reef fish radically alter their patterns and colours when they reach maturity. In their younger years, fish

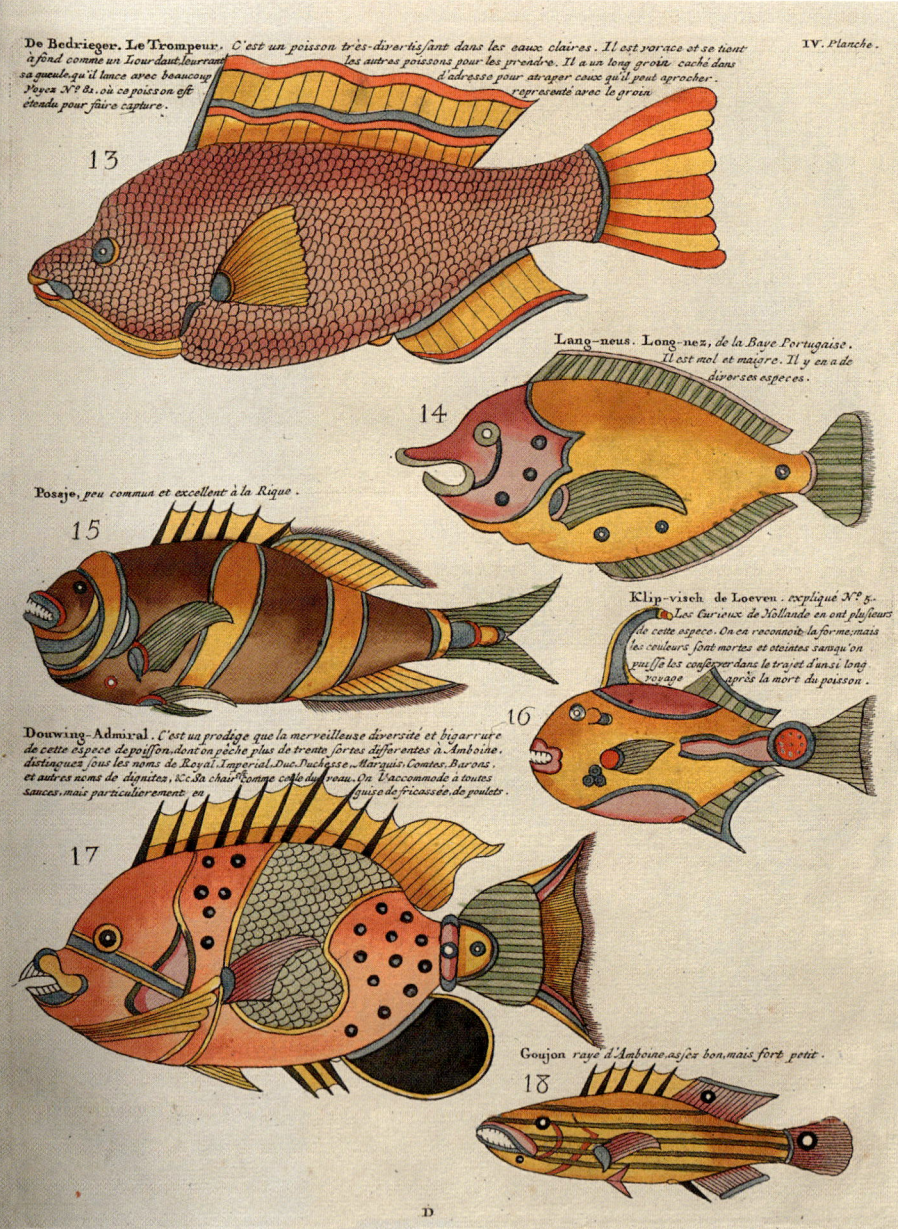

De Bedrieger. Le Trompeur. C'est un poisson très-divertissant dans les eaux claires. Il est vorace et se tient à fond comme un Lourdaut, lourrant les autres poissons pour les prendre. Il a un long groin caché dans sa queule, qu'il lance avec beaucoup d'adresse pour atraper ceux qu'il peut aprocher. Voyez N° 82. où ce poisson est représenté avec le groin étendu pour faire capture.

IV. Planche.

13

Lang-neus. Long-nez, de la Baye Portugaise. Il est mol et maigre. Il y en a de diverses especes.

14

Posaje, peu commun et excellent à la Rique.

15

Klip-visch de Loeven. expliqué N° 5. Les Curieux de Hollande en ont plusieurs de cette espece. On en reconnoit la forme; mais les couleurs sont mortes et eteintes tanqu'on puisse les conserver dans le trajet d'un si long voyage après la mort du poisson.

16

Douwing-Admiral. C'est un prodige que la merveilleuze diversité et bigarrure de cette espece de poisson, dont on pêche plus de trente sortes differentes à Amboine, distinguez sous les noms de Royal. Imperial. Duc. Duchesse. Marquis. Contes. Barons et autres noms de dignites. &c. Sa chair comme celle du veau. On l'accommode à toutes sances, mais particulierement en guise de fricassée, de poulets.

17

Goujon rayé d'Amboine, assez bon, mais fort petit.

18

D

aren't looking to find and control a territory. Their juvenile colours signal their non-threatening stage of life, making them less likely to get chased off and attacked by older fish.

Unfamiliar Sharks

As people were getting to know better the inhabitants of shallow seas and artworks were shifting from symbolic approximations towards real, living species, there were animals that were a challenge to accurately draw because they were rarely seen whole and intact. Basking sharks (*Cetorhinus maximus*) are the second-largest fish in the ocean, after the tropical whale sharks, and can grow to 10 or even 15 metres (33 or 49 ft) long; most would just fit inside a London double-decker bus with the seats stripped out. Hauled out of the water by fisheries that used to target their oily livers, the bodies of these huge animals collapsed and looked nothing like their living selves. Seen alive, they are quite un-shark-like in character and rather strange-looking things. They aren't swift predators that chase after prey, but gentle, slow-cruising filter feeders that swim near the surface of shallow seas, their huge mouths open to sift rich blooms of minute plankton with their gill rakers. An old name for them was the broad-headed gazer.

Many artworks depict a side-on view of basking sharks with mouths wide open, sometimes with astonished wide-eyed gazes, as in the book *A History of the Fishes of the British Islands* by Jonathan Couch. Another image in that book shows an endearing male basking shark, with spots on his snout (not accurate) and a pair of pink testicle-like appendages – the male sharks' functional equivalent of a penis, known as claspers – which basking sharks do possess but that rarely so obviously dangle behind the living animal. A basking shark in the mid-nineteenth-century *Natural History of New York* is a little more true to life and again drawn

Illustrations of fish by Samuel Fallours, including majestic angelfish (bottom left) and triggerfish (middle right), from Louis Renard, Poissons, écrevisses et crabes . . . (1718), vol. II.

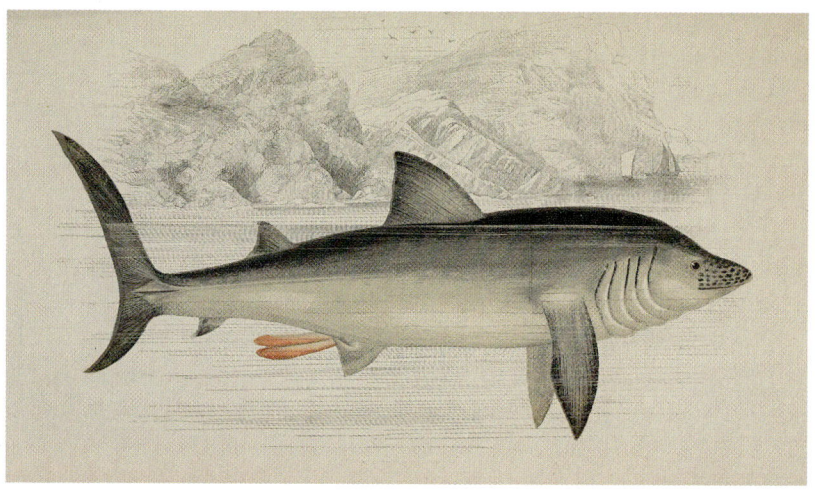

side-on. I've found no historic drawings of basking sharks that show them head-on, as many modern underwater photographs do, letting the viewer gaze into the cavernous mouth and the arched gill rakers inside. Presumably few people had jumped in the water and seen a basking shark swimming towards them to witness this planktonic perspective on these very large fish.

Small-bodied species of shark living in shallow waters have been known in Europe since medieval times and were generally referred to as dogfish or *cazón*. Many of these animals are at odds with the stereotypical view of sharks being huge, grey, torpedo-shaped predators that must keep swimming in order to breathe. It's true that sharks lack the gas-filled swim bladders of bony fish and their bodies are negatively buoyant, so they sink if they stop swimming. And some can only ventilate their gills by opening their mouths as they swim and letting the water pour in. However, plenty of other sharks quite peacefully rest on the seabed and actively pump water into their mouths.

Drawings of many of these shallow-water sharks feature in a monograph by the influential German scientist Johannes Müller. In the 1830s, Müller developed a new system for classifying sharks and in 1841 he published *Systematische Beschreibung der Plagiostomen*

(Systematic Description of Plagiostomen) – *Plagiostomen* being Müller's term for sharks and their close relatives. Among the beautiful illustrations are sharks that have evolved unusual ways of hiding themselves on the seabed. The Japanese wobbegong (*Orectolobus japonicus*, labelled *Crossorhinus barbatus* by Müller) lives in the western Pacific, a habitat that ranges from the waters of Japan and Korea to Vietnam and the Philippines, where their dark mottled patterns and whiskery fringe around the head act as camouflage. From a similar region, the brownbanded bamboo shark (*Chiloscyllium punctatum*) lies still on the seabed, its outline broken up by wide bands of colour. The puffadder shyshark (*Scyllium edwardsii*, later renamed *Haploblepharus edwardsii*) from South African waters gets its common name from the orange bands that resemble those of puff adders, and from the sharks' defence strategy: when these timid, seabed-dwelling sharks feel

Opposite: *Basking shark, illustration from Jonathan Couch,* A History of the Fishes of the British Islands, *vol. 1 (1868 edn).* Below: *Basking shark, illustration by Arthur Bartholomew, from* Frederick McCoy, Prodromus of the Zoology of Victoria, *vol. 11 (1890).*

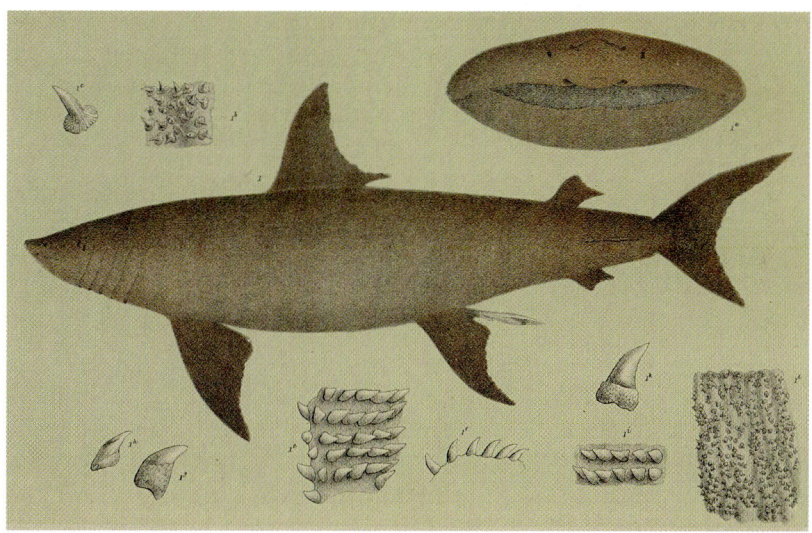

threatened they curl their bodies into a tight circle and cover their eyes with their tail, thus protecting the most vulnerable part of their bodies.

As Müller knew well, sharks have some curious-looking cousins, like the shortnose guitarfish (*Rhinobatus brevirostris*, renamed *Zapteryx brevirostris*) and the chola guitarfish (*Rhinobatus undulatus*, now *Pseudobatos percellens*), both from the southwest Atlantic between Brazil and Argentina. These animals look rather like a cross between a shark and a stingray that's been stretched out to a point at the front end. Müller's work also features more circular stingrays, like the striped panray (*Platyrhinus schoenleinii*, now *Zanobatus schoenleinii*) from West Africa, with distinct markings across its back; the sepia stingray (*Urolophus aurantiacus*) from Japan and the East China Sea, depicted in bold orange; and the Chinese fanray (*Platyrhina sinensis*), shown with delicate pink fins and a characteristic constellation of white thorns dotted across the back.

Stunning colour illustrations of species from all around the world swim through the pages of what remains probably the most

Below: *Japanese wobbegong, illustration from J. Müller and J. Henle,* Systematische Beschreibung der Plagiostomen *(1841).*
Opposite: *Shortnose guitarfish, illustration from J. Müller and J. Henle,* Systematische Beschreibung der Plagiostomen *(1841).*

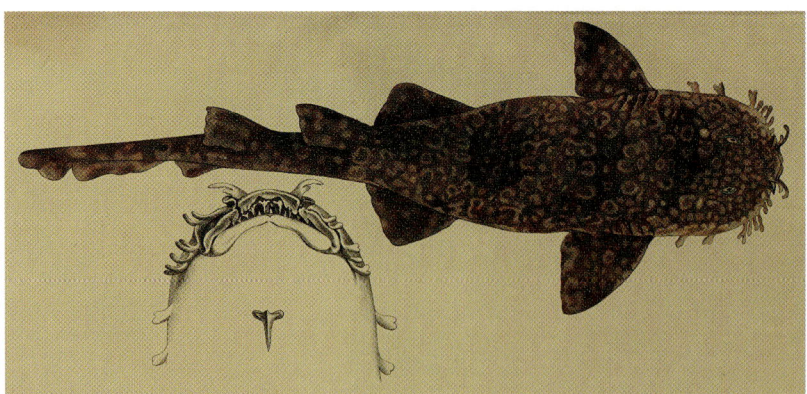

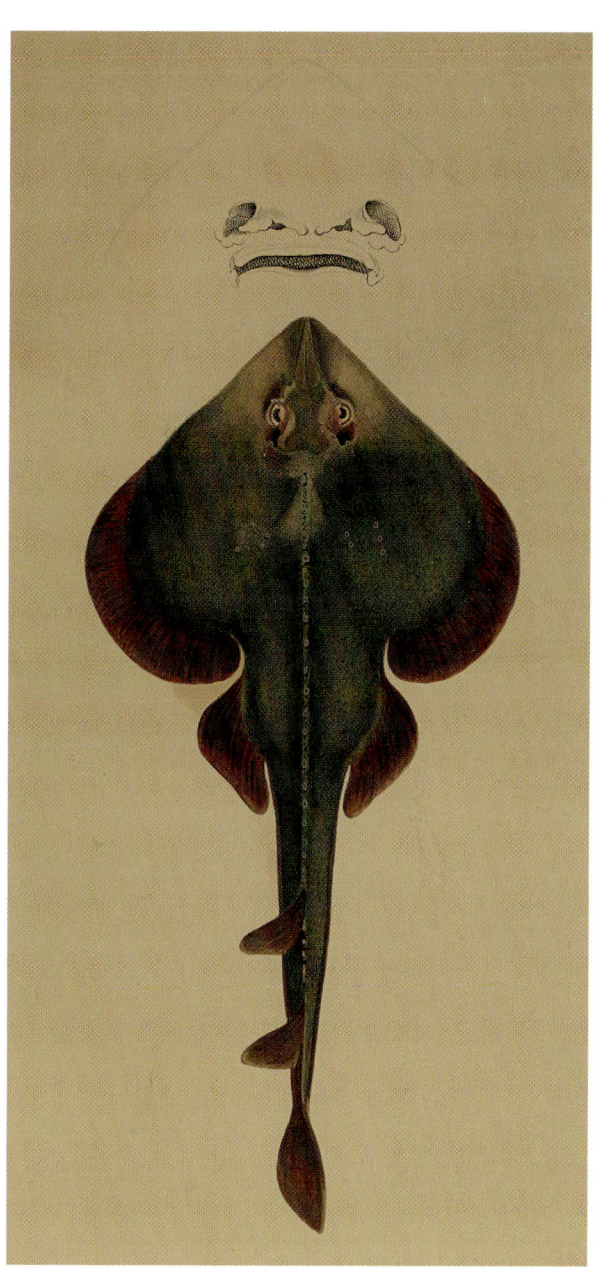

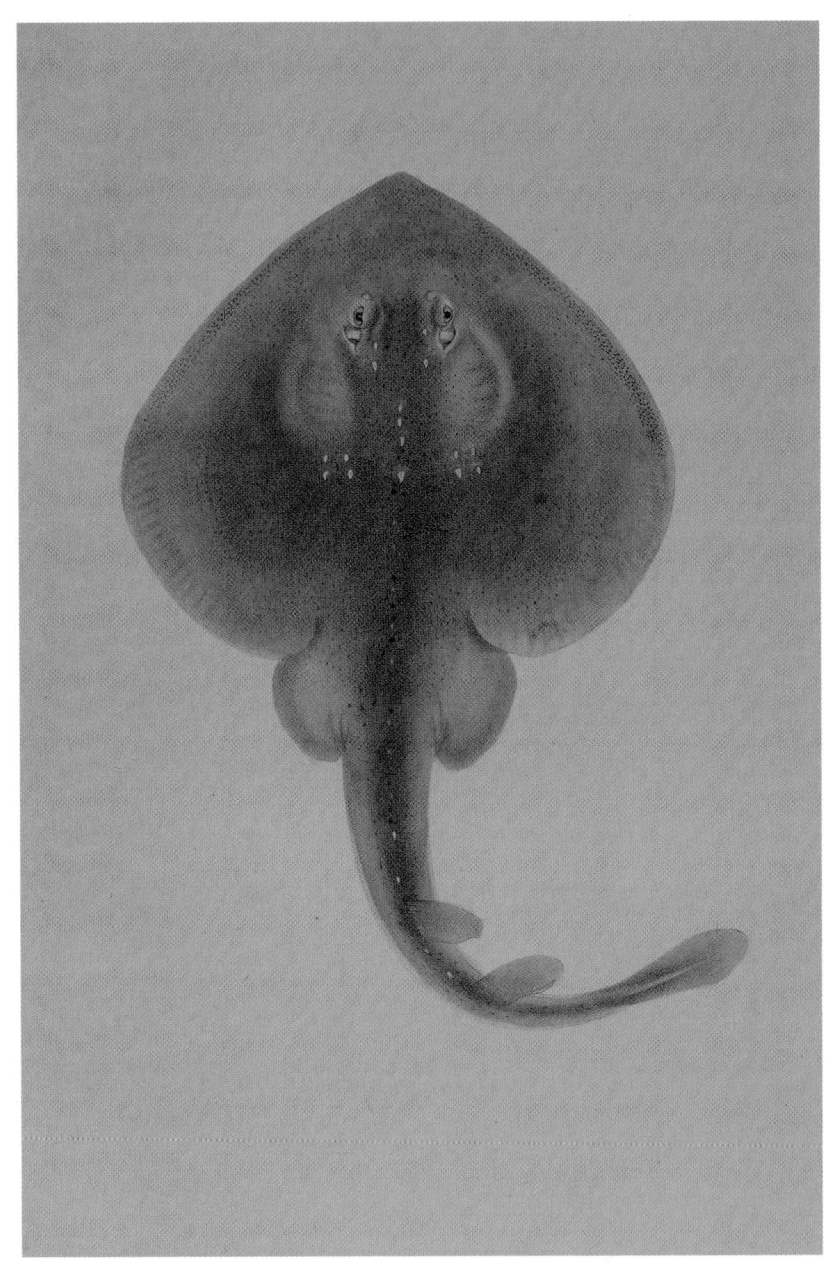

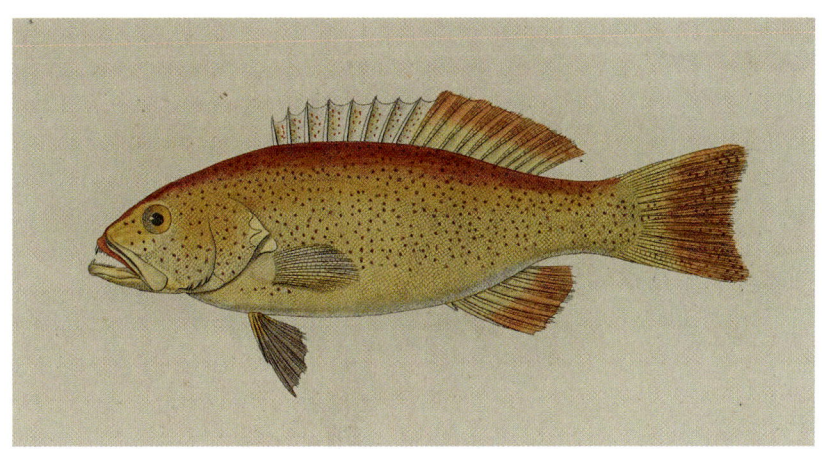

Opposite: *Chinese fanray, illustration from J. Müller and J. Henle,* Systematische Beschreibung der Plagiostomen *(1841).* Above: *Leopard coral grouper (*Plectropomus leopardus*), from Georges Cuvier,* Histoire naturelle des poissons, *vol. II (1828).* Below: *Semicircle angelfish (*Holacanthus semicirculatus, *now* Pomacanthus semicirculatus*), from Georges Cuvier,* Histoire naturelle des poissons, *vol. VII (1831).*

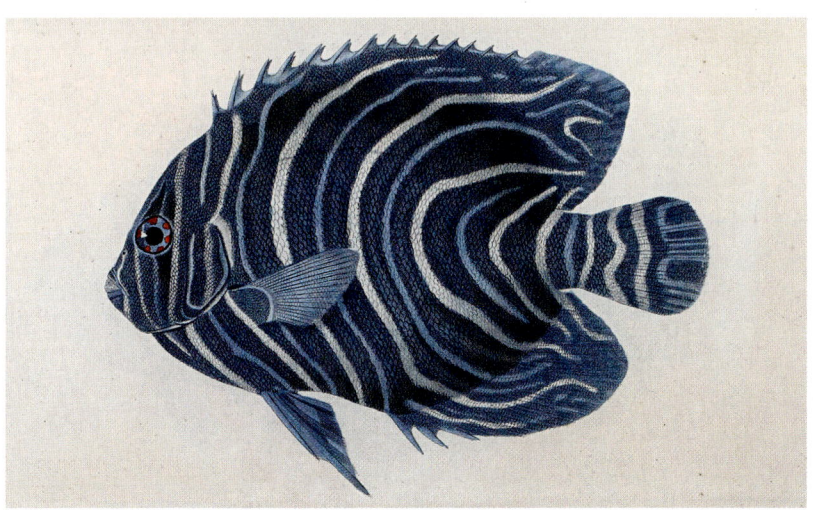

complete and detailed illustrated scientific book of fish. It took more than forty years and multiple authors to produce the 24-volume series *Histoires naturelles des poissons* (Natural Histories of Fish).

The work was started by French scientist Georges Cuvier, from the Museum of Natural History in Paris, who is remembered most as a pioneering palaeontologist who advanced the highly contentious idea that ancient species have died out and gone extinct. He also had a burning ambition to create a compendium of all the world's fish. Collectors stationed across the globe sent preserved specimens to Cuvier, which he then described and named, and commissioned illustrations of – thousands of them were species new to science.

After his death in 1832, Cuvier's fish work was continued by his student, Achille Valenciennes, and on his death he passed the baton to another French scientist, Auguste Duméril. When Duméril died in 1870, Cuvier's dream of publishing a book of all known fish was still unfinished, showing just how monumental a task it would be. It is unlikely that anyone would decide to try this today because nowadays scientists tend to focus on particular groups of fish, not all of them all at once. But if they did, the compiler of a complete book of fish would have more than 30,000 scientifically known species to get through.

Putting Fish in Their World

The fish imagery that adorned natural-history books and illustrated catalogues from the sixteenth to the nineteenth century were largely intended as academic works that classified and named the rich diversity of fish, and helped professional and amateur naturalists identify the species they were finding. And they all had one thing in common: they all depicted fish in isolation. Each illustration, from the thick-lined woodcuts to the meticulous, naturalistic engravings, showed an individual fish floating in empty space (and more often than not facing towards the left side of the page).

Outside the European scientific establishment, other artistic traditions have favoured placing fish in the context of their surroundings. Fish are often portrayed in the process of getting caught by fishers, including many generic fish of no particular type that lend themselves to being symbols of lost souls being rescued from a sea of sin. Others appear as specific species, as on various North American cigarette cards from the turn of the twentieth century. These collectable trading cards issued by tobacco companies to advertise their brands and stiffen their cigarette packets commonly showed sportsmen and actresses, and some series featured wild animals. The 'Fishers and Fish' series released in 1888 by W. Duke, Sons & Co. showed female anglers posing

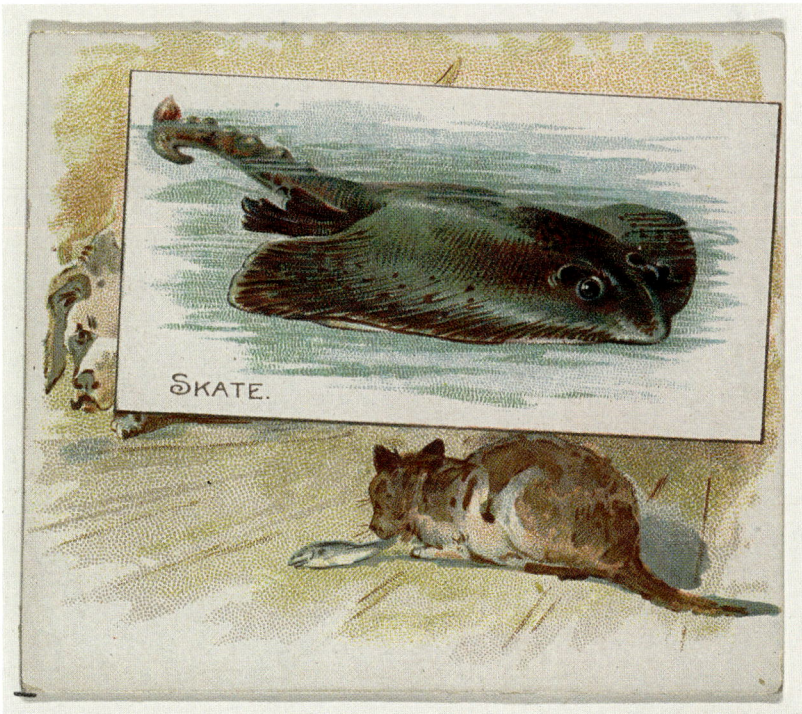

Skate, cigarette card from the 'Fish from American Waters' series issued by Allen & Ginter, 1889.

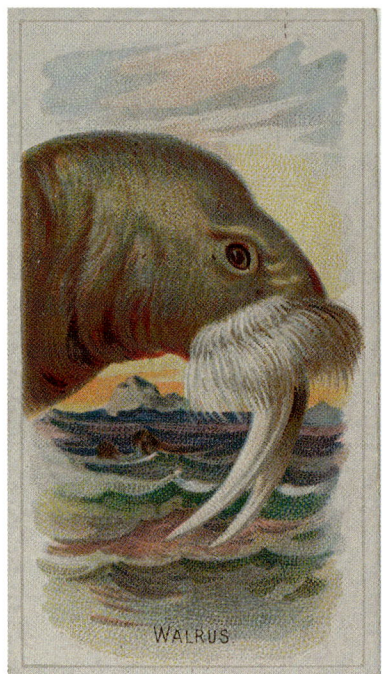 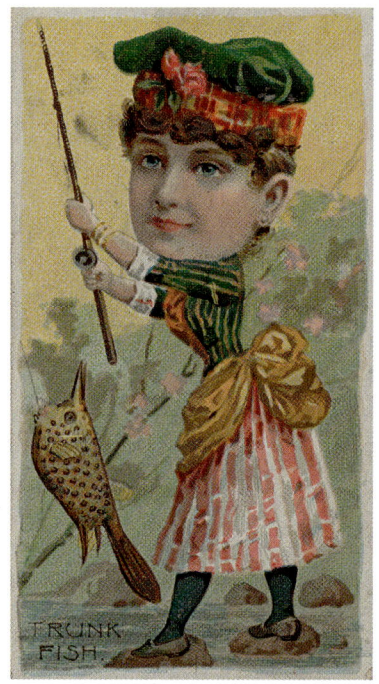

Left: *Walrus, cigarette card from the 'Wild Animals of the World'
series issued by Allen & Ginter, 1888.* Right: *Angling woman with
trunk fish, cigarette card from the 'Fishers and Fish' series issued by
W. Duke Sons & Co., 1888.*

with their catches, elegantly dressed in different outfits and with
disproportionately large heads. Fish were evidently popular
among smokers at the time. Another cigarette brand, Allen &
Ginter, released a series in 1889, 'Fish from American Waters',
featuring dozens of species alongside fishing boats and different
types of fishing gear.

Native habitats for countless illustrated fish are market
stalls, butcher's blocks and dining platters. Fish are regarded not
so obviously as food – but are nevertheless long dead – in many
pictures that show them fished out of the sea and lying on a beach,
with a maritime scene of some sort going on in the background.

Dutch printmaker Albert Flamen produced many such works; two goggle-eyed lumpsucker fish lie in the foreground with a sailing boat at work fishing in the distance; a pair of plaice covered in spots lie stiffly together; a trio of sole; a pile of skates. As well-observed as images of this kind were – there's no mistaking the cod with barbels dangling from their chins, or the claspers on the underside of a catshark – of greater interest to the ecologically minded are scenes that illustrate fish swimming in their own wild world.

Japanese art has a long tradition of depicting living rather than dead fish, often placing these animate creatures within underwater scenes. Nature was a common subject for artists who portrayed other living things not as part of a human-centric world, but inhabiting their own harmonious spaces.

It is fitting, although not intentional, that many fish appear in the genre of Japanese art, popular from the seventeenth century through to the nineteenth, called *ukiyo-e*, meaning 'pictures of a floating world'. The 'floating world' in question referred to hedonistic districts in the city of Edo (now Tokyo), which from the early 1600s was experiencing an economic boom. A burgeoning working class had enough disposable income to spend on attending kabuki theatre and buying paintings and colour prints to decorate their homes.

Elegant courtesans and actors were fashionable early subjects in *ukiyo-e*, and it was only later that landscapes and nature became more desired. An iconic work from this era is Katsushika Hokusai's oceanic scene *The Great Wave off Kanagawa* from 1831. That same year, the prolific printmaker produced a still-life of a flatfish, a scorpionfish and a scattering of shells, with a branch of bamboo as a hint of vegetation behind them. Similar still-life prints of fish, plants and flowers were produced by Utagawa Hiroshige, another master of late *ukiyo-e*. Hiroshige produced more naturalistic underwater scenes featuring fish interacting elegantly with their surroundings. In a print dating from the 1830s, a shoal of five trout slip across the scene between blue shaded lines that allude to the underwater currents of the river as the fish swim towards the sea.

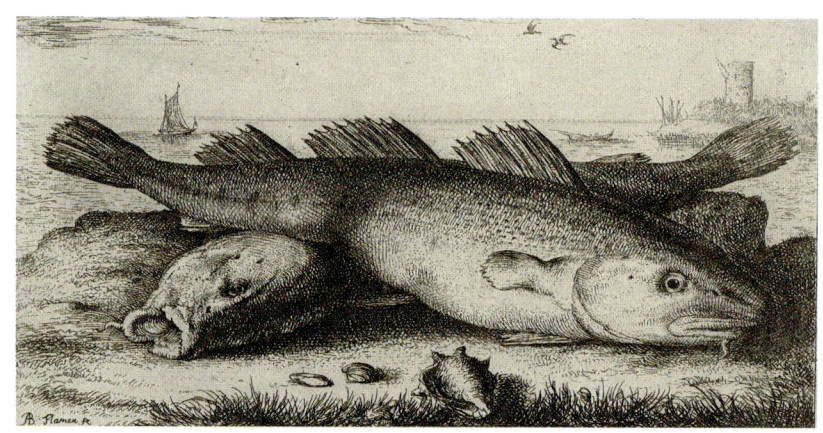

Above: *Albert Flamen*, Two Cod on the Beach, *1664, etching*. Below: *Albert Flamen*, Two Plaice on the Beach, *1664, etching*. Opposite above: *Utagawa Hiroshige*, Fugu and Inada Fish, *from the series 'Uozukushi' (Every Variety of Fish), 1840s, colour woodblock print*. Opposite below: *Utagawa Hiroshige*, Trout, *c. 1832–3, colour woodblock print*.

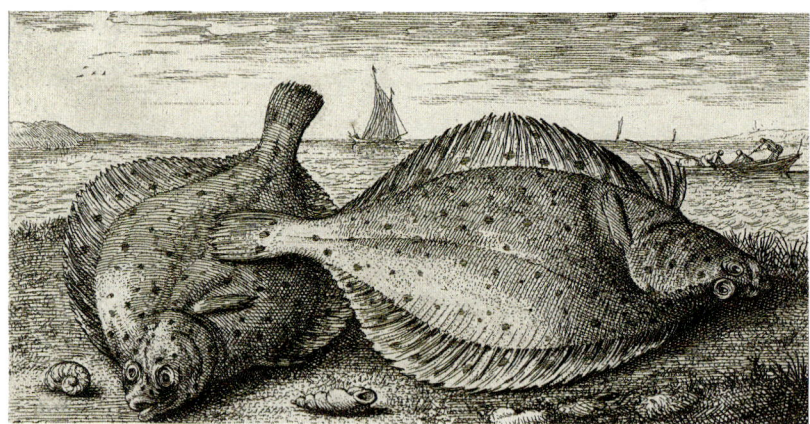

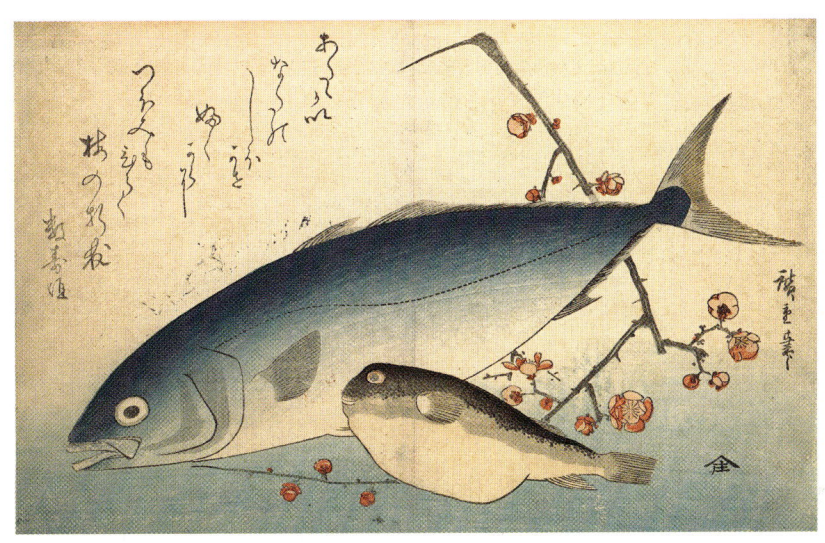

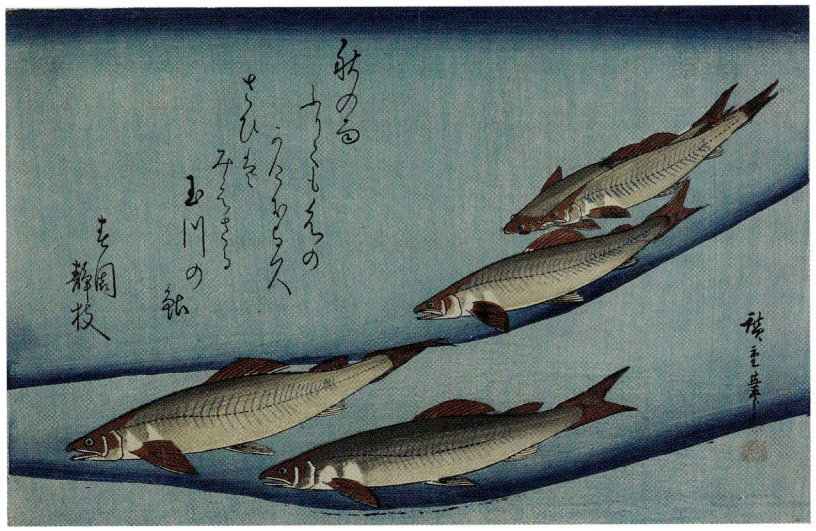

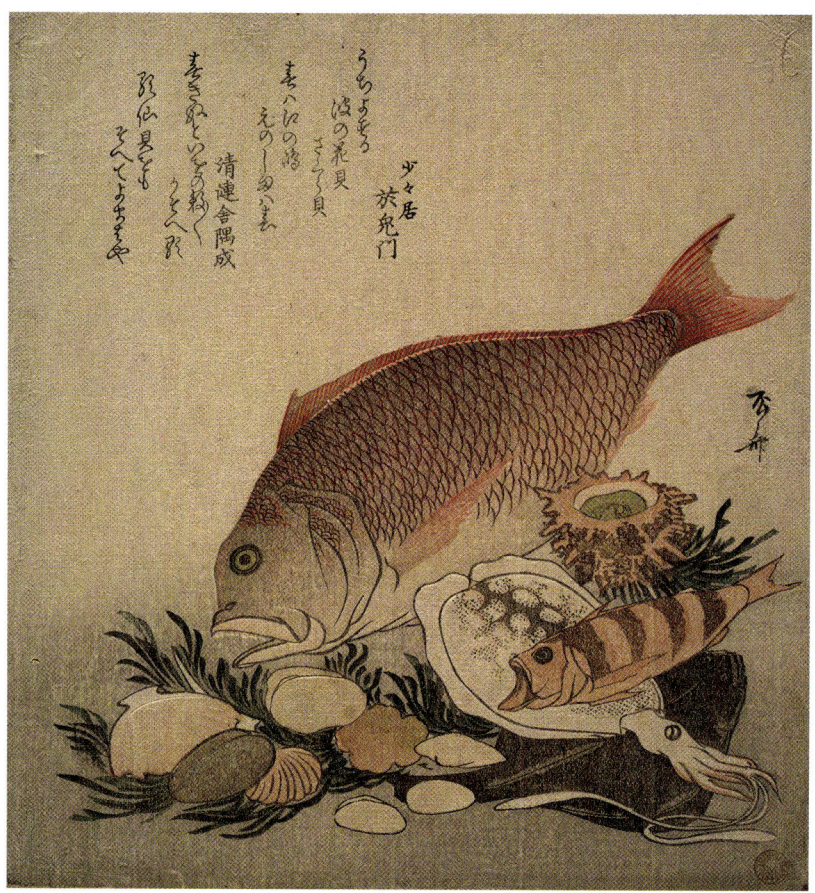

Among the works of a student of Hokusai, Ryūryūkyo Shinsai, is a rare underwater scene showing a mix of fish species, a cuttlefish and seashells nestled in seaweed. Several decades later, in the 1890s, Seki Shūkō produced a series of fishes swimming through the water that seems to ripple across the page, with a sprig of aquatic vegetation here and there, indicating that the bottom of the sea is there too.

Aquatic scenes also appear in Japanese metalwork and carvings from a similar time. *Tsuba* are circular metal hand guards that were made to be both functional parts of swords, placed between the hilt and the base of the blade, and also portable, often double-sided, works of art. Among the animals and naturalistic scenes is a *tsuba* with an icefish entwined in waterweeds.

Handles for small knives, *kozuka*, also featured aquatic creatures and their habitats – an eel slithering past a clump of intricate vegetation, a pufferfish with a few golden fronds of seaweed and seashells, a needlefish pushing its slender snout through the water.

Meanwhile, Japanese recreational fishers developed their own traditions for depicting fish. *Gyotaku* (meaning 'fish impression' or 'fish rubbing') began in the 1830s as a way of accurately recording the size of a fisher's trophy catches, an equivalent of having prize fish stuffed and mounted. The technique involves carefully covering the outside of a fish in ink (sometimes while it's still alive), then pressing a sheet of paper onto it to leave an intricate inked impression. The fish can then be cleaned of the non-toxic ink, sold and eaten, and the living ones let go. Since fishers started out recording their catches this way, *gyotaku* has flourished as an art form, with a recent resurgence of interest and people around the world learning the technique. Painters can add details on the fish, arrange them together in a shoal and fill in a scene behind for them to swim through.

There is scientific value to these traditional works of art, due to the custom of writing the location and date of the fish's capture on the page, as well as the type of fishing gear used to catch it, all useful information for tracking historic fish populations. For a 2020 study, a team of Japanese scientists collected *gyotaku*

Opposite: *Ryūryūkyo Shinsai,* Large and Small Fish Swimming Among Shells and Moss at the Bottom of the Sea, c. *1830, colour woodblock print.* Below: *Knife handle (*kozuka*) with eel, Japanese, c. 1615–1868, copper alloy and gold.*

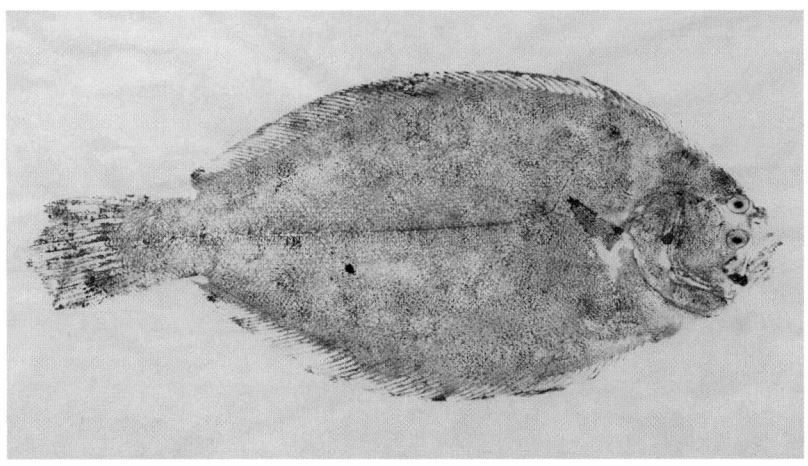

Above: *Flatfish* gyotaku *(inked fish print)*. Opposite above: *Nazca ceramic bowl with crab, 2nd–4th century*. Opposite below: *Nazca ceramic cup with rows of lobsters or crayfish, 180* BC–AD *500*.

images from different regions of Japan and showed that they faithfully reflect the abundance of species in the wild. Out of several hundred *gyotaku* fish, seven from the island of Hokkaido were the critically endangered Japanese huchen (*Hucho perryi*), an ancient migratory species of salmon, and three from Miyazaki Prefecture were Japanese barramundi, or *akame* (*Lates japonicus*), which are at risk of going extinct. The *gyotaku* prints open a window into the past, revealing where fish were living and getting caught in times before every angler had a smartphone in their pocket to snap photos.

Spineless Creatures of the Shallows

Besides fish, the shallow seas are also inhabited by a great variety of invertebrates. These animals all lack bones and internal skeletons and either have entirely soft, supple bodies or keep their squashable innards inside a hard covering, or exoskeleton, such as the seashell-making molluscs. Crustaceans, in particular crabs,

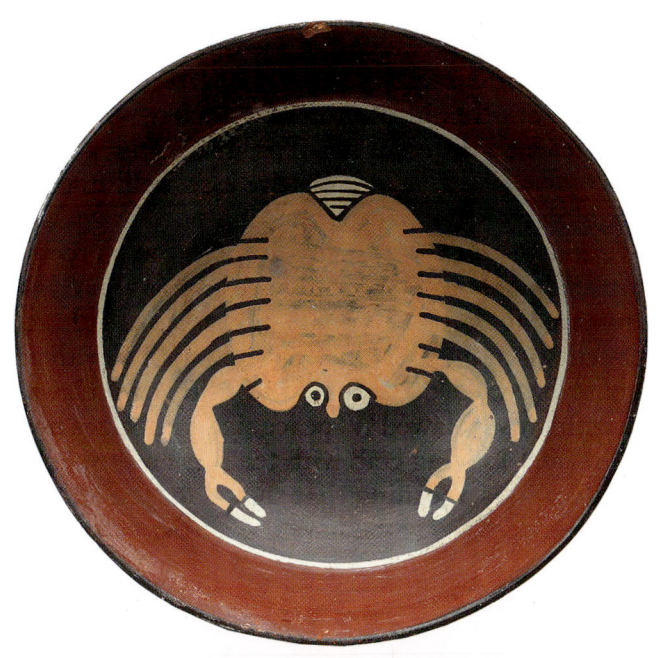

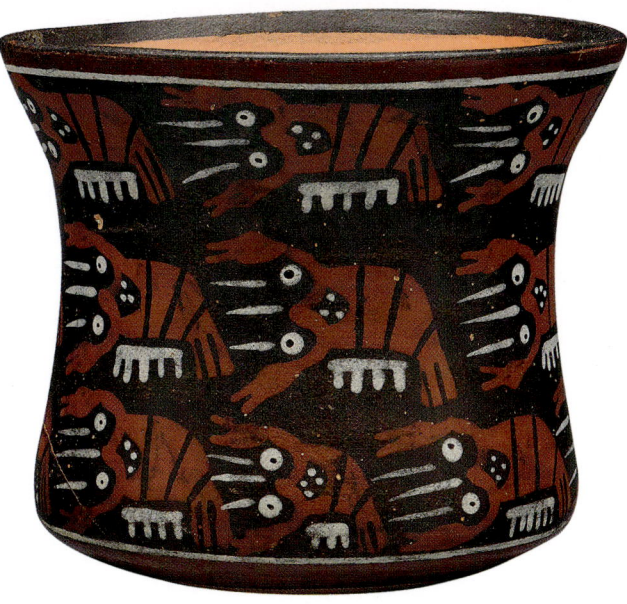

shrimp and lobsters, are a common sight marching across the shallow seabed or darting through the water with flips of their tails. They are another important and ancient source of human food and present a different set of animal features compared to fish: their powerful claws, inquisitive eyes held high on stalks, their tough carapace that needs cracking open to get at the meat inside. Various aspects of spineless animals can be seen creeping across artworks and objects, motifs and designs around the world.

Crustaceans were well known in the Nazca culture of Peru. A pottery cup from the first millennium is painted with a repeated pattern of stylized red lobsters, all holding up their claws is if they are performing two-clawed 'V' peace signs. A Nazca bowl shows a rather unusual-looking crab with ten legs and two claws; most crabs are decapods, meaning that they have ten appendages in total. This one does, however, have the correct anatomical detail

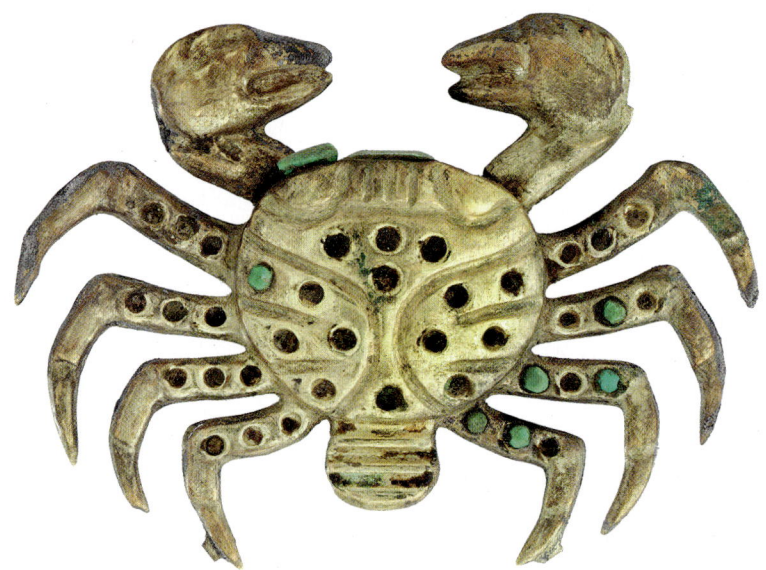

Above: *Moche gold crab with green inlay, 6th–7th century.* Opposite above: *Nazca ceramic lobster effigy vessel, 4th–6th century.* Opposite below: *Moche gold and silver nose ornament with shrimp, 6th–7th century.*

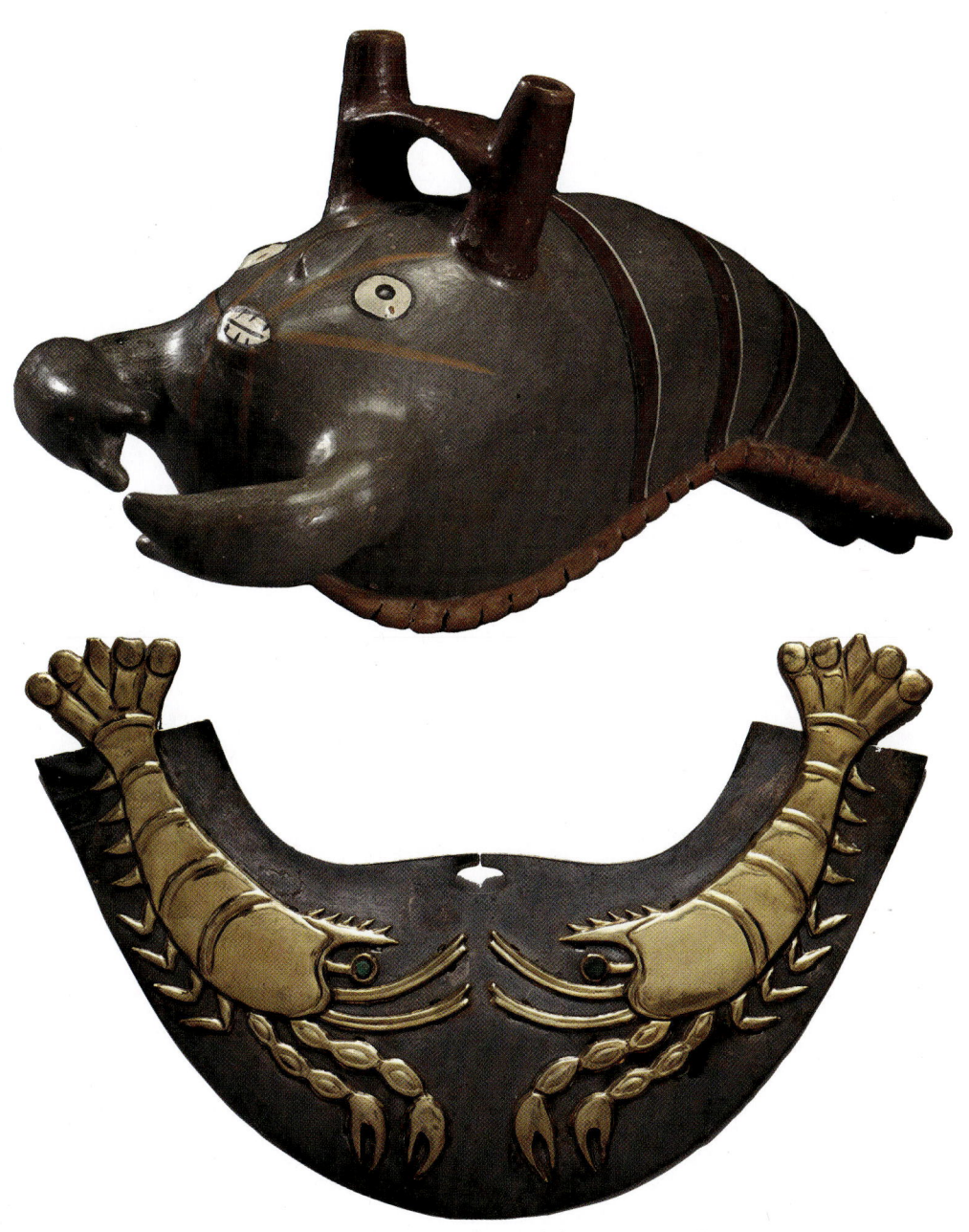

of the ridged flap of abdomen folded under the body. Female crabs brood their eggs under this flap before they hatch and crawl or swim away.

Hundreds of years later, that same crustacean feature appears in a gold crab with inlaid precious stones from the Moche civilization of northern Peru. This crab also has the conventional complement of four pairs of legs plus a pair of claws. Crabs clearly played a significant role for Moche people, as shown by the splendid iconography they produced. Some ceramics and metalwork show whole and intricately observed animals, as on a gilded copper disc with a crab at its centre surrounded by concentric rings of fish. This crab has embossed eye stalks, lifelike claws and articulated points of the crab's jointed legs – another key feature of arthropods (meaning 'jointed legs', Arthropoda is the phylum of invertebrates including crustaceans, insects, millipedes and spiders). Decorative ceramics show crab–human hybrid warriors. Many pots were painted with a fearsome man, his body a crab's, with eight crab legs plus two human legs, brandishing a terrifying pair of serrated claws. Some crab deities had a huge human face grafted onto the crab's carapace.

Other crustaceans that appear in Moche ceramics and metalwork are lobsters and crayfish, which, like crabs, were a major source of food. A stunning gold and silver nose ornament depicts two shrimp, possibly a freshwater species from the rivers of Peru, gazing at each other with green stones for eyes. A sharp, spiny rostrum sticks forwards from the front of each head, a distinguishing part of shrimp anatomy which these animals would have had in real life and used for defence and fighting.

Moche people may have revered crustaceans because of their powerful claw weapons or perhaps the crabs' supernatural ability to effortlessly span the boundary between land and sea. Even though they creep about the shoreline, and some species spend much of their lives on dry land, crabs can never separate themselves entirely from water. They carry water in their gill chamber so they can keep breathing on land, but they must return to the sea to lay their eggs.

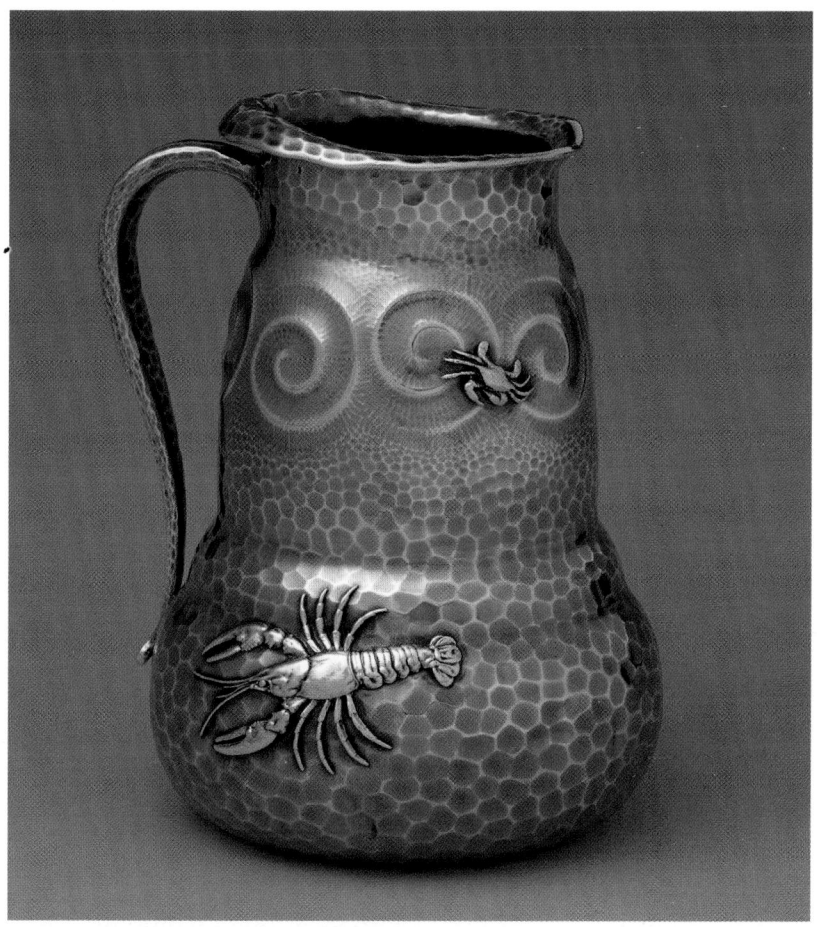

Silver pitcher with lobster and crab, Tiffany & Co., c. 1880.

Shell-making conch and spiny oysters weren't the only molluscs rendered in Moche motifs. The class of living molluscs known as cephalopods (meaning 'head foot') encompasses the octopuses, squid and cuttlefish. Except for the chambered nautiluses, their ancestors stopped making shells millions of years ago, leaving them with unprotected soft bodies. Most octopus species live close to the seabed and many in shallow seas, where

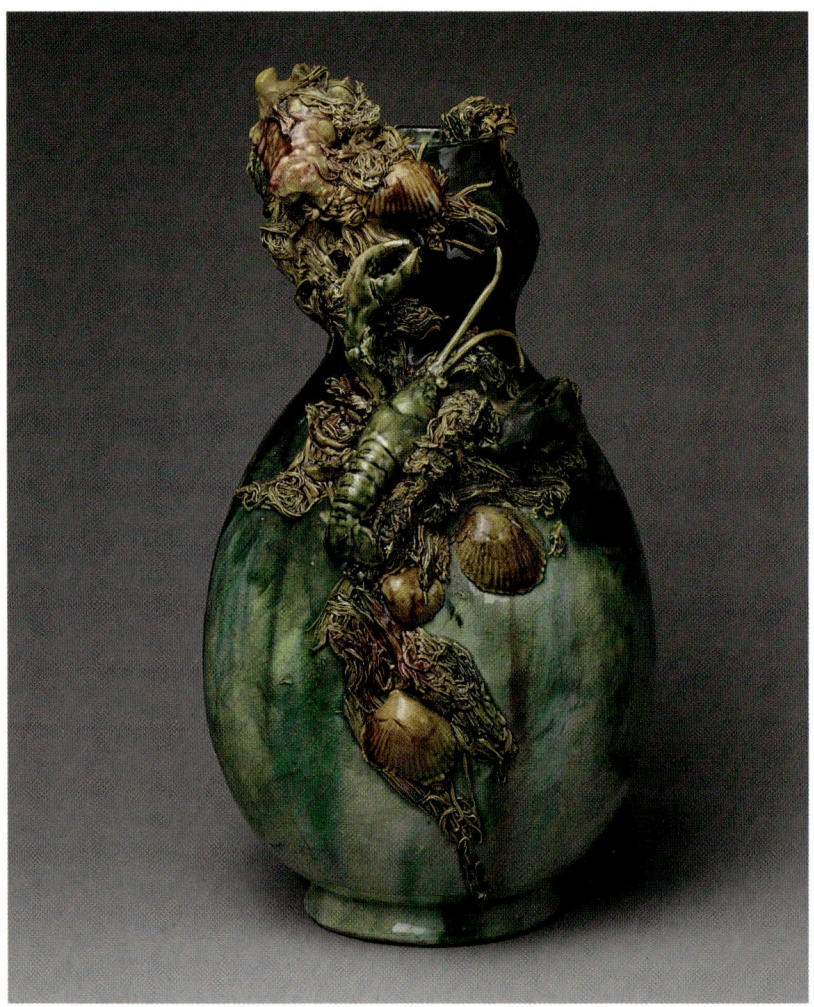

people have hunted them since ancient times, often via the simple method of leaving pots for these secretive animals to climb into and hide.

The eight distinctive arms of octopuses appear in stunning gold Moche ornaments, known as frontlets. Ceramic figurines show elite people, perhaps religious leaders, seated cross-legged and wearing an eight-armed headdress tied with a band around their foreheads. The decoration's arms are arranged with a neat spiralling curve at each tip – something living octopuses do – and a fringe of triangles where the octopus's suckers would be. In the centre of the headpiece is a grimacing, fang-toothed face, perhaps an image of the Moche deity *Ai Apaec*.

Jumping to the modern era, crustaceans and cephalopods continue to appear in precious objects and artworks. In the 1870s and '80s, the American jewellery design house Tiffany & Co. produced a range of innovative and exquisite silverware. A silver pitcher and a chocolate pot, both designed by Charles Osborne, feature lifelike casts of lobsters and crabs crawling across them. A ceramic vase from the same period by Thomas J. Wheatley of Cincinnati, Ohio, is made in the French technique called barbotine for adding three-dimensional decorations to pottery. Wheatley's vase is beautifully encrusted in sea life, with cockleshells, a sweep of seaweed and a splendid lobster climbing to the top.

Opposite: *Ceramic vase with marine life (lobster, shells and seaweed), by Thomas J. Wheatley, 1882.* Below: *Salvador Dalí, Lobster Telephone, 1938, steel, plaster, rubber, resin and paper.*

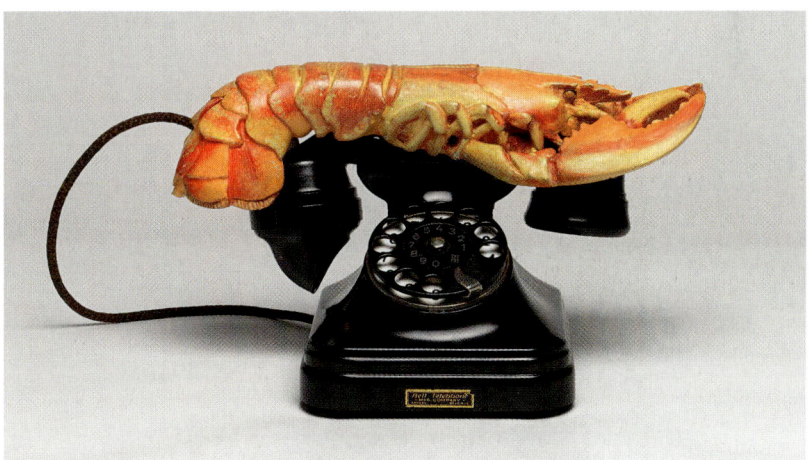

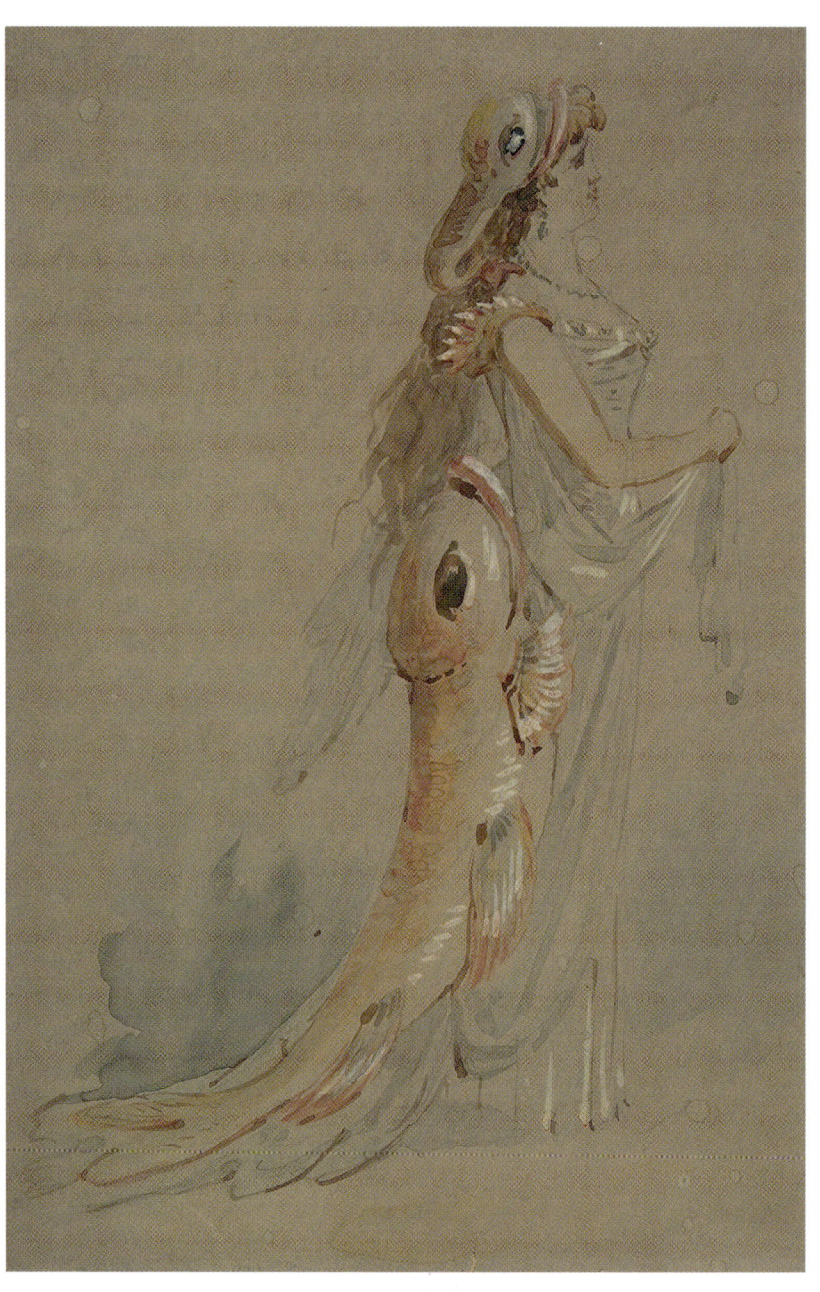

Lobsters took on a whole new meaning in works by Spanish surrealist Salvador Dalí. Inspired perhaps by nineteenth-century Parisian writer and poet Gérard de Nerval, who kept a pet lobster named Thibault, in 1935, in the magazine *American Weekly*, Dalí published a drawing called 'Man Finds a Lobster Instead of a Phone'. The cartoon shows a man looking in horror as he realizes the telephone in his hand is in fact a large-clawed crustacean. In New York in 1936 Dalí displayed a live lobster perched on a red telephone. A similar arrangement in Paris in 1938 he called the *Aphrodisiac Telephone*. For Dalí, lobsters were highly sexually charged animals. Made into a telephone handset, the speaker is required to talk into the underside of the lobster's tail, where the female carries her clutch of fertilized eggs. These installations

Opposite: Anonymous drawing of a woman wearing a costume with fish motifs, second half of the 19th century. Below: 'Falmouth Borough Octopus attempting to grasp the parishes of Falmouth and Budock', political map by Edwin T. Olver, c. 1882.

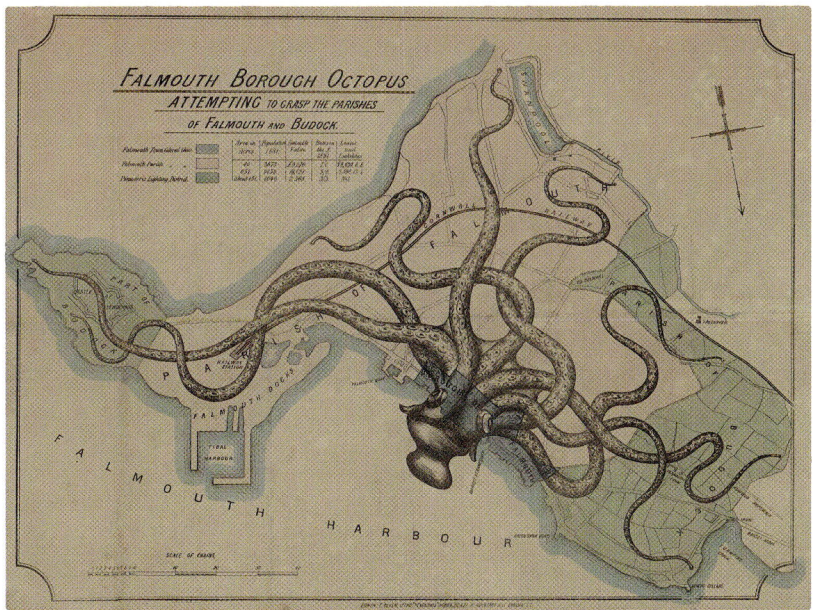

inspired Dalí and his friend and patron, the wealthy British poet and art collector Edward James, to produce a series of fully working lobster telephones. Dalí helped to give James's London home a Surrealist makeover, including their cherry-red sofas shaped as the Hollywood actress Mae West's lips. Eleven plaster lobster telephones were modelled on real lobsters, seven painted white and four red, and they were hollowed out to fit the working parts of the telephones.

The lobster theme returned in a collaboration between Dalí and Italian fashion designer Elsa Schiaparelli. The A-line evening gown from 1937 is made from sheer ivory silk organza and has a red lobster appliqué across it, plus a few sprigs of seaweed here and there; it reportedly took six people 250 hours to craft the dress. That same year, American socialite Wallis Simpson was photographed by Cecil Beaton for *Vogue* magazine wearing the lobster dress. She stands in the dappled sunlight of a wooded glade, holding bunches of flowers, the lobster visible in the delicate folds of the dress, the tail positioned as Dalí provocatively intended, between her legs.

Meanwhile, images of octopuses have taken on a more sinister persona. Satirical cartoons and propaganda posters have used these eight-armed molluscs as a trope to represent power grabbing by politicians and dictators. Russia is depicted as a bloated octopus in a war map of Europe by British graphic artist Fred Rose, first published in March 1877, two months after Russia invaded the Ottoman Empire. A poster from 1885 from a British seaside town shows the 'Falmouth Borough Octopus attempting to grasp the parishes of Falmouth and Budock'. Sprawled across the map is a beady-eyed octopus and its eight entwining arms.

Returning to the Japanese Edo period, octopuses, lobsters and other invertebrates were common characters. Three little crabs stand by the water's edge in an 1830 *surimono* woodblock by Yashima Gakutei, another of Hokusai's pupils. *Surimono* (meaning 'printed things') are especially lavish woodblock prints that used the finest paper and pigments, and were less well known than *ukiyo-e*. They weren't intended for wide distribution and were

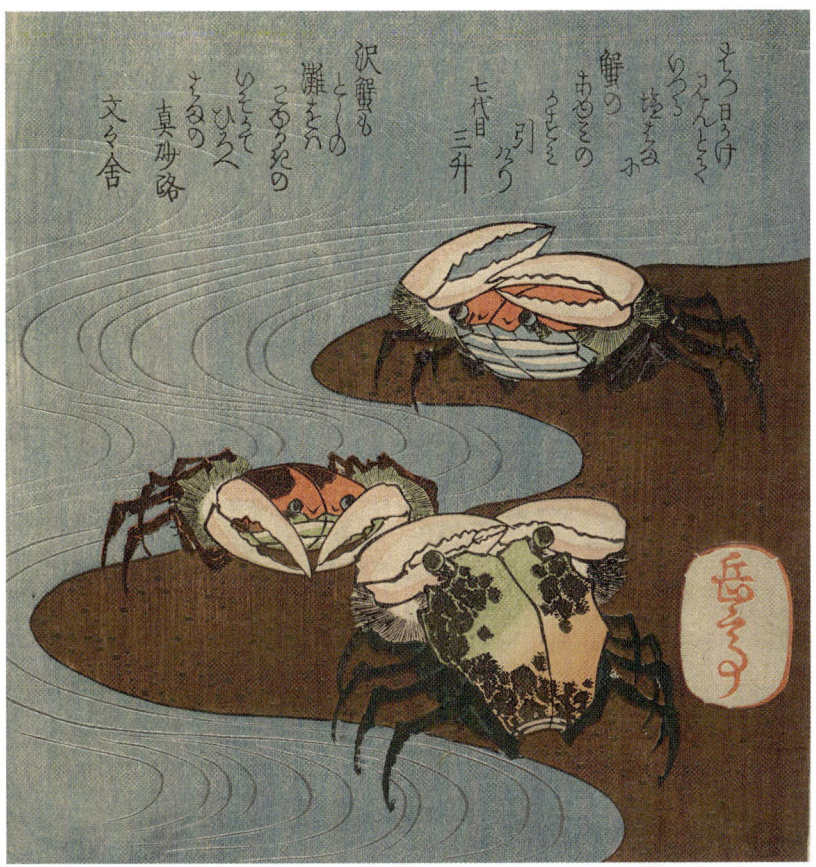

Yashima Gakutei, Three Crabs at Water's Edge, c. *1830, colour woodblock print (*surimono).

rarely put on sale, but privately produced in small numbers to give to friends and family as greetings cards at New Year.

Surimono were often combined with poems, which helped to explain the often puzzling artworks. One such piece depicts a crab, sitting on a black lacquered hat worn at court, and a vase of flowers, all arranged on a decorated cloth. The meaning of the piece becomes clear in lines of the poem referring to the coastal city of Akama, where the major sea battle of Dan-no-ura took place in 1185. The fleet of the Genji clan defeated the Heike clan, in a major turning point in the history of Japan, as the ruling class of emperors and aristocrats was overtaken by shoguns and warriors. Akama is home to a native Japanese crab known as *heikegani* (*Heikeopsis japonica*) with carapaces marked in patterns that many have said resemble human faces. Traditional beliefs hold that the *heikegani* crabs are reincarnations of the defeated Heike warriors.

Heikegani crabs got a mention in American astronomer Carl Sagan's 1980 television series *Cosmos* as an example of unintentional human-driven selection. He said that when fishers caught crabs with faces on their shells, they would throw them back alive, not wanting to kill and eat the warrior's offspring. The more humanlike the crab, the greater their chances of surviving and reproducing, expanding the population of crabs with face patterns. Thus, over time, the crabs became even more humanlike. While this is fine in theory, a snag is that *heikegani* crabs are not a species targeted and caught in fisheries.

Crabs were exquisitely carved in *netsuke* and *ojime*, the small toggles and beads that were essential items in Edo-period Japanese fashion. The style of kimono worn by men had no pockets or sealed sleeves, which the women's versions had, so they carried tobacco, medicines and other items in small containers suspended on cords

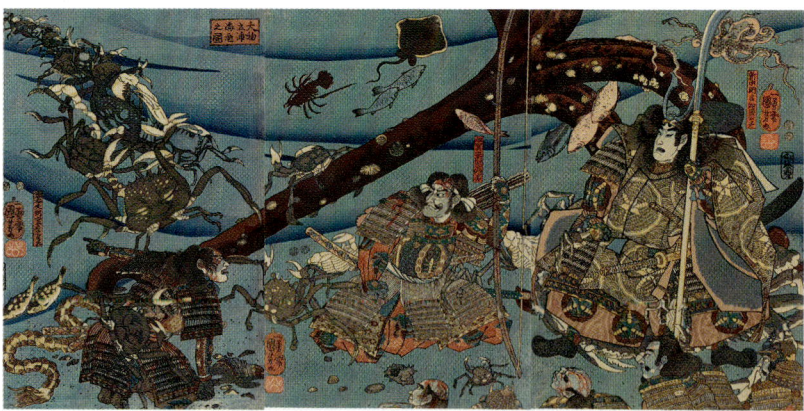

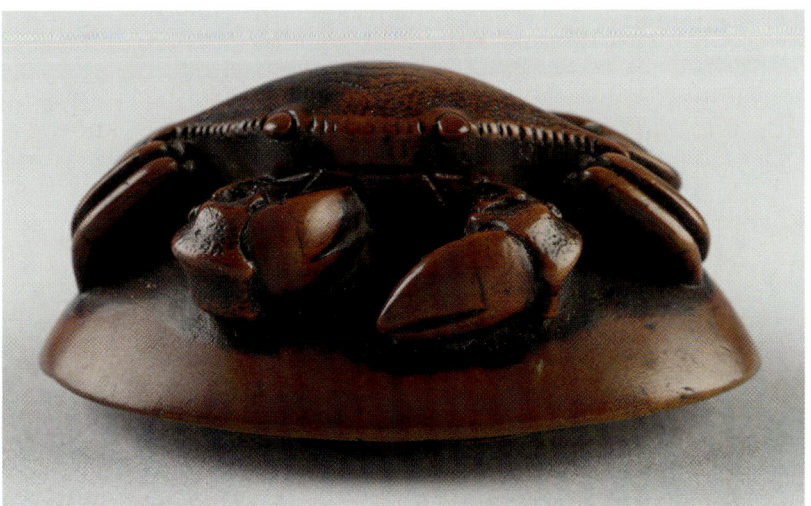

Opposite: Heikegani *crabs with human-like faces in battle: Utagawa Kuniyoshi*, At the Bottom of the Sea in Daimotsu Bay, *1851–2, oban triptych, colour woodblock print.* Above: Netsuke *of crab on a shell, 19th century, wood.*

looped over their kimono sash. *Netsuke* were counterweights to the containers and *ojime* could be slid along the cord to open or close the container. Similar to the sword guards and knife handles, *netsuke* and *ojime* were not just practical items but tiny works of art – miniature sculptures just a few centimetres in size, commonly carved from ivory or lacquered boxwood – and they have become highly sought-after collectors' items. *Netsuke* crabs in museum collections are full of personality and intricately observed from life. One sits atop a clam-shell, raises a claw to its mouth to feed and looks ready to scurry away.

Other crustacean varieties also make their appearance in Japanese art and artefacts. A print by Seki Shūkō depicts a pair of hermit crabs, with their second-hand shells left behind by sea snails after they died. A spiny lobster reaches its long antennae across the front of a *tsuba* and gazes out with a pair of beady gold eyes.

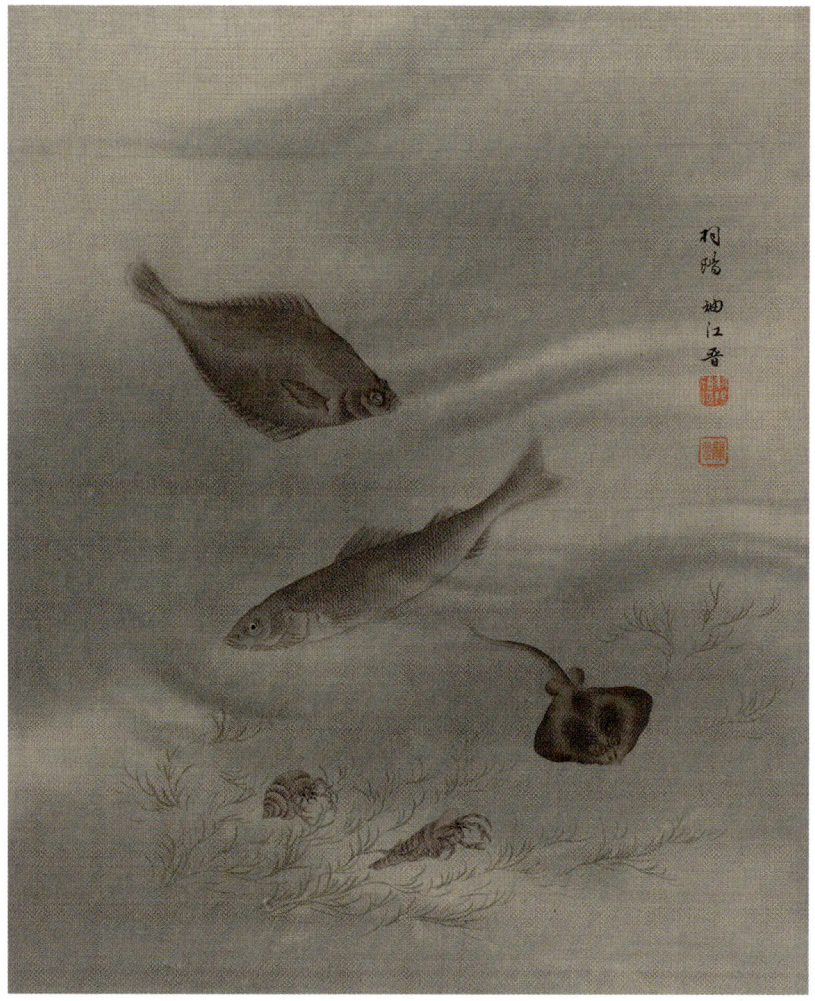

Octopuses, squid and cuttlefish weave their arms and tentacles all through Japanese traditional arts, in *gyotaku*, *ukiyo-e*, *surimono*, *tsuba* and *netsuke*. These cephalopods have long been important species in coastal fisheries and cuisine around Japan, which is no doubt why they are such prominent figures in arts and crafts. They appear as compelling characters, cute and cartoonish,

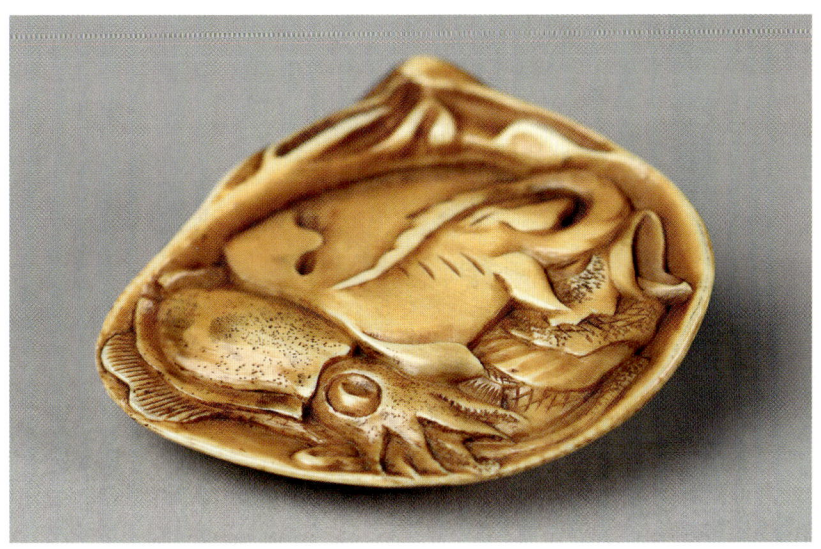

Opposite: *Seki Shūkō*, Fishes *(flatfish, sting ray and hermit crabs)*, c. *1890–92, ink and colour on silk.* Above: Netsuke *of a shell filled with cuttlefish and other sea life, 19th century, ivory.* Below: Netsuke *with a mix of sea life by Ikkosai Toun, mid-19th century, ivory.*

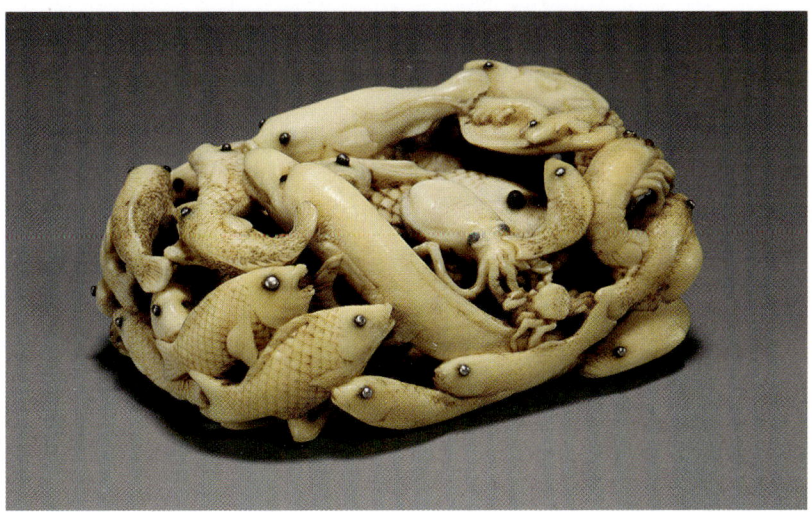

terrifying or racy; they attack boats, fight people and sometimes pleasure fishermen's wives.

Japanese art also contains many examples of creatures from the shallow seas that don't feature by themselves, but are arranged together with a mix of other species: a *netsuke* with octopuses, skates, seashells and fish all bundled up together in the small, miniature sculpture; a *tsuba* with a catfish and a lobster hanging out

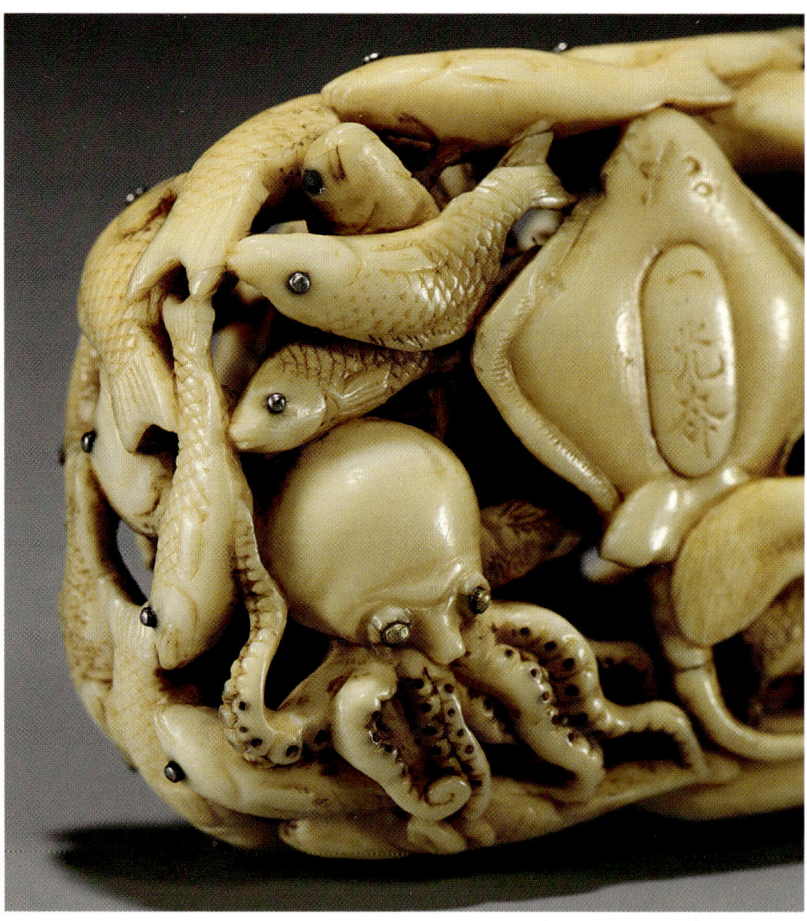

Netsuke *with a mix of sea life by Ikkosai Toun, detail.*

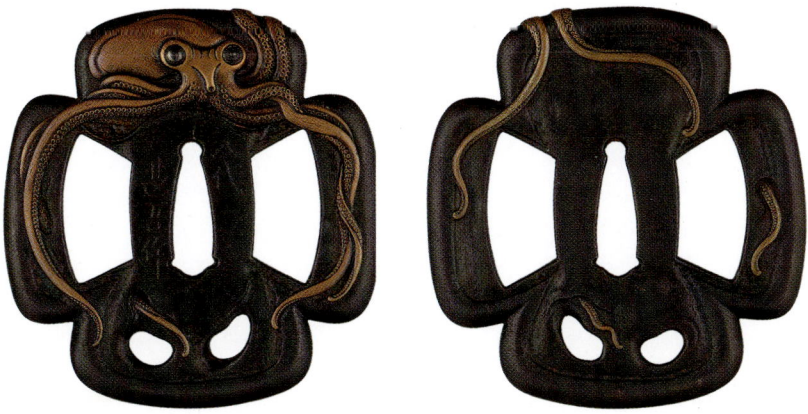

*Sword guard (*tsuba*) with octopus, Japanese, late 18th century, iron and copper, front and back.*

together beneath the waves; another with a flatfish, cuttlefish and fish swimming in swirling spirals of water. These microcosms are glimpses of ecosystems and they lead us towards the tangled webs of interaction that take place between marine species in the real, wild world.

Fish Eats Fish

Creatures from shallow seas were often assembled together in ceramics and mosaics made by ancient cultures surrounding the Mediterranean. In the Bronze Age Minoan civilization based on the island of Crete, pots for holding oils and ointments were often covered in sea life. Known as the Marine Style, these works featured dolphins, fish, seashells, sea urchins and seaweed, along with what I contend are some of the most charismatic ceramic octopuses ever created, with mesmerizing round eyes and arms bending all over the place. Later cultures, such as the Mycenaeans, were inspired by the Minoan ceramics to create their own stylized octopus vases.

Many of the sea creatures in artworks in the Graeco-Roman world were likely grouped together for mainly decorative effect, although people were obviously aware that these species occur together in nature. A hint of a more ecological way of thinking, and of animals actually interacting with each other, can be seen in a troupe of marine animals that appears in a well-known mosaic from the Roman city of Pompeii. Most are identifiable as particular species thanks to the fine detail in the mosaic work: a torpedo ray with large spots on its disc-shaped body; a pair of cat-sharks with elongated, sleepy-looking eyes; a gurnard with its distinctive, fanlike pectoral fins. Most of the fish pay each other little attention and are merely occupying the same space, except

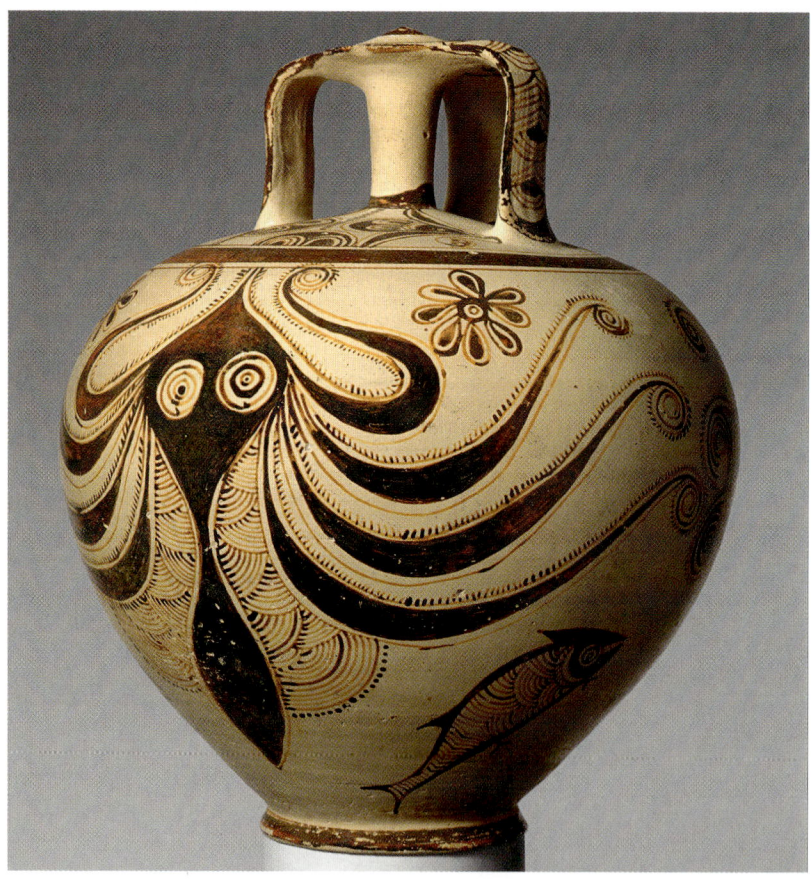

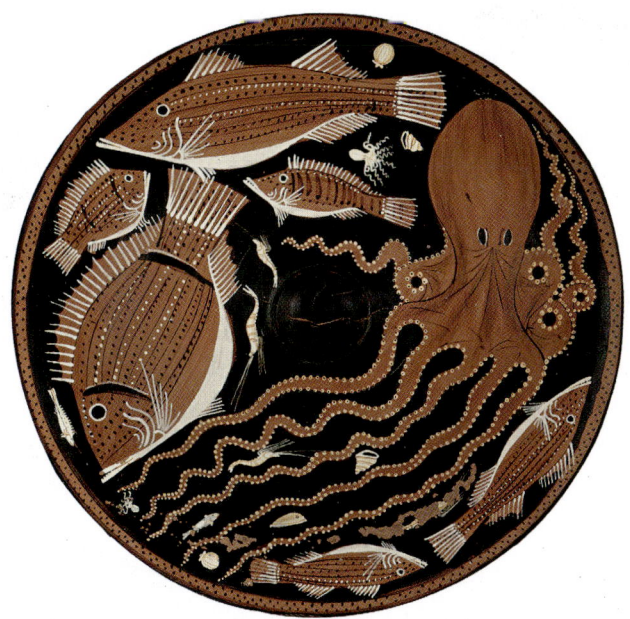

Opposite: *Mycenaean stirrup jar with octopus*, c. *1200–1100* BC, terracotta. Above: *Red-figure fish plate (octopus, mullet, bream, shellfish)*, *Paestan*, c. *340–30* BC, *ceramic*.

for a pair of animals that draws the eye to the centre of the piece. An octopus fixes its two round eyes on the viewer, while its arms are entwined around the body of a lobster that's disproportionately large compared to the other species in the mosaic. Lobsters are a common prey species for octopuses. The soft molluscs camouflage themselves on the seabed, changing the colour and surface texture of their skin to match their surroundings. Then they pounce and, as the mosaic depicts, grapple their target in their eight muscly arms before delivering a lethal, poison-laden bite with their bird-like beak.

Moving ahead several centuries, one of the most famous scenes depicting a mix of interacting fish, and undoubtedly one of the most bizarre, is Pieter Bruegel the Elder's 1556 drawing *Big Fish Eat Little Fish*. A giant fish lies dead on the shore

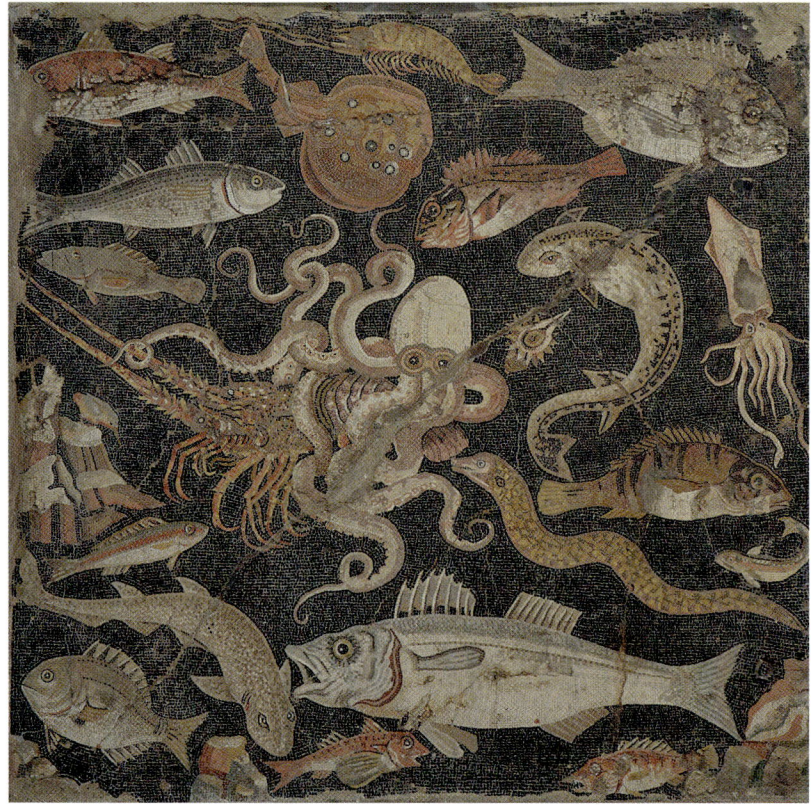

and tumbling from its huge mouth into the shallows is a torrent of other fish, including some with smaller fish stuffed in their mouths. More fish are emerging from the big fish's belly, where a man is slicing it open with a giant butter knife.

The fish-within-fish theme is repeated around the scene; a trio of fish in the water are busy eating each other like a set of Russian dolls; a fisher sitting in a boat is pulling one fish from another like an ichthyological magician.

Pieter van der Heyden's 1557 engraved version of the piece spells out the proverbial message, reminding the viewer that big fish will always eat smaller fish. In other words, the world is full of injustice and the little guys will always get brutally beaten down by the rich and powerful. There are nods to the strange imaginings of Hieronymus Bosch, who died several decades previously, in the fantastical fish flying high through the air and another fish plodding off on a pair of human legs.

Strange as this scene is, looking at it from a biologist's perspective, this is a food web of sorts. It shows the kind of study that scientists still undertake to understand who eats whom in an ecosystem. While there are more sophisticated tools available and modern ways to analyse chemical traces that reveal dietary habits, there is still important information to be gained from slicing open a dead fish's belly and seeing what's inside.

Creating an ecological message was presumably not Italian painter Giuseppe Arcimboldo's intention when, in 1566, he painted

Opposite: *Mosaic of marine life, Pompeii*, c. *100* BC. Below: *Pieter van der Heyden, after Pieter Bruegel the Elder,* Big Fish Eat Little Fish, *1557, engraving.*

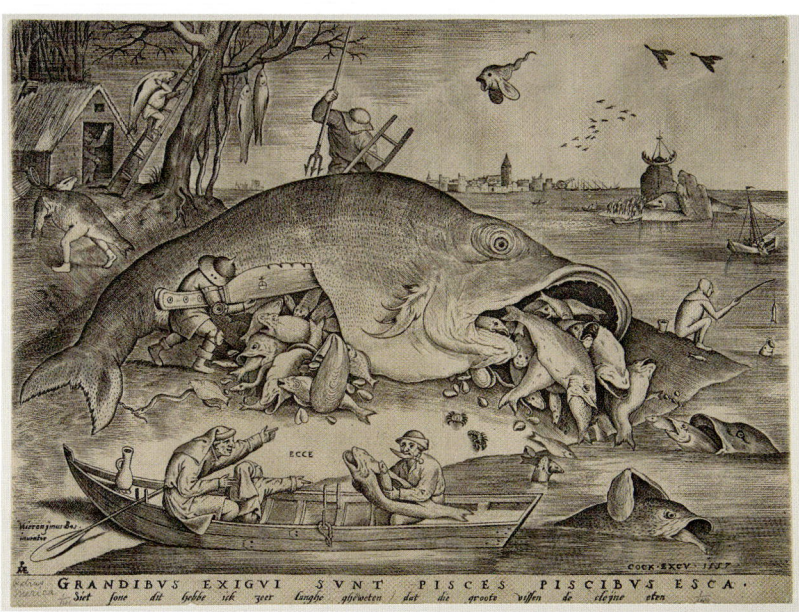

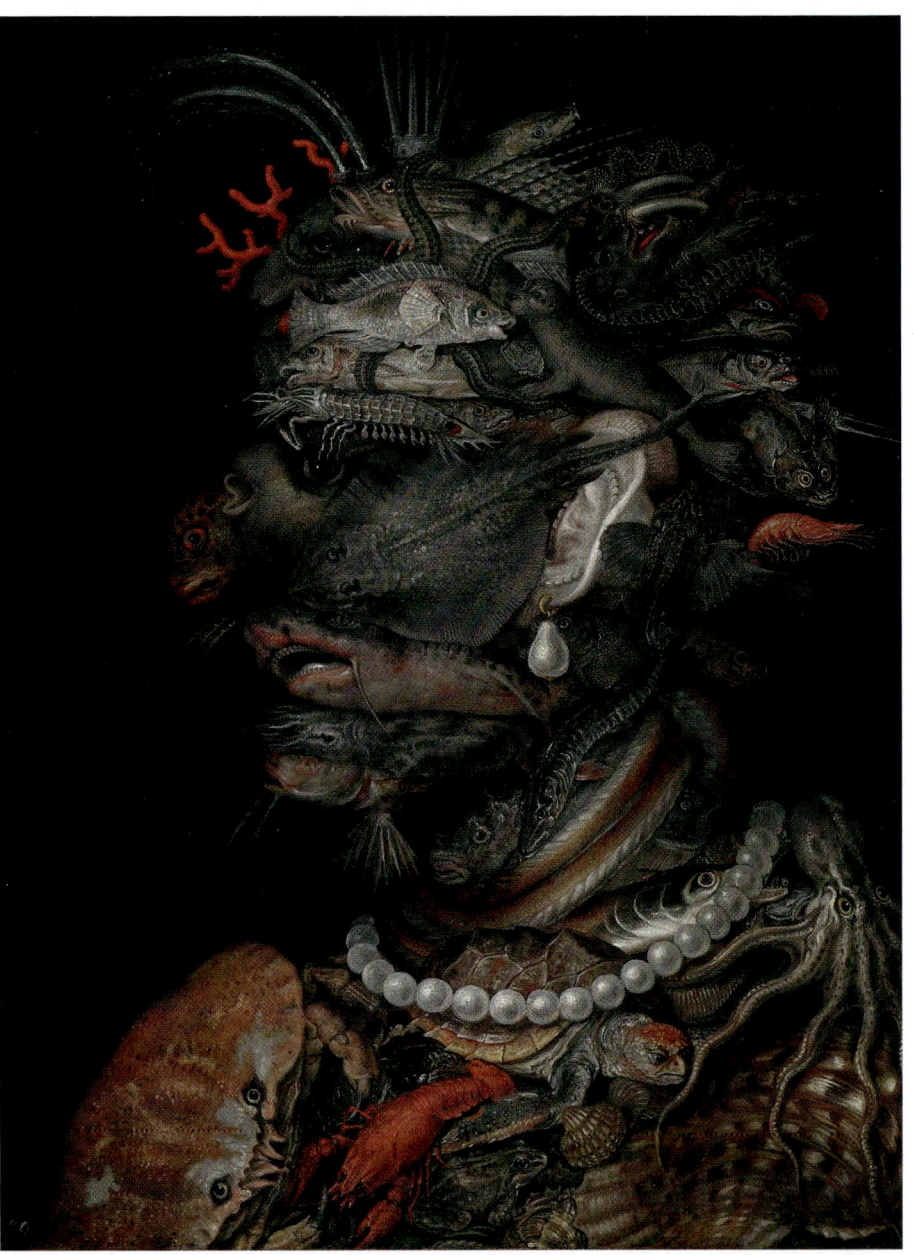

a human head composed of a diverse array of fish species. The fish ensemble represented 'Water' in his *Four Elements* series, alongside a head made of birds for 'Air'; another comprising sheep, lions, elephants and deer for 'Earth'; and a fourth head with his hair on fire. Even so, Arcimboldo's fish head is his own interesting ecosystem. The man's eye belongs to a pufferfish and his mouth is a catshark; a goatfish provides the man's short beard; his ear is the underside of a skate (pierced with a pearl earring). Piled into his hair and clothing there's a seahorse, a sturgeon, a wrasse and an eel.

Other sea creatures are here too, as they should be; after all, the water is not just a home for fish. Crustaceans and cephalopods take their places: a large crab on the man's chest, a bobtail squid and a sharp-clawed mantis shrimp, an octopus and a lobster (although cooked, by the look of the red carapace – most lobsters are blue when they're alive and only change colour when they are boiled and dead). There's also a sea turtle, a walrus and a seal. And while these species would not necessarily all live together in real life, Arcimboldo nevertheless displays in his own strange way the living riches of the ocean.

Quite a different work of *Four Elements* was created over the following two decades by the Dutch artist Joris Hoefnagel. Born into a wealthy merchant family in Antwerp, he travelled widely in Europe in his youth and later fled the Netherlands when his family fortune was plundered during the Eighty Years War with Spain. Hoefnagel was hired as a court artist by various dukes and emperors in Munich, Frankfurt and Vienna. Among his employers was the Holy Roman Emperor Rudolf II, who commissioned Hoefnagel's illuminated manuscripts to accompany the works of art and nature he collected in his *Kunstkammer*, or cabinet of curiosities. Rudolf's collections also came to include Hoefnagel's *Four Elements*.

Eventually spanning four volumes, Hoefnagel's book contains hundreds of miniature paintings depicting thousands of species representing the four elements: insects (fire), quadrupeds

Giuseppe Arcimboldo, Water, *1566, oil on panel.*

and reptiles (earth), birds (air) and aquatic animals and shells (water).

A copy of the water volume *Animalia aqvatilia et cochiliata (Aqva)*, dated to between 1575 and 1580, is held at the National Gallery of Art in New York and is rarely put on public display. A digital archive of the paintings, a mix of watercolour and gouache, shows how exquisite they are in their size and detail. Each one is a small underwater or coastal scene, painted within a golden oval on a page the size of a postcard. Most have a waterline horizon, roughly two-thirds of the way up. Within these enclosed pools of water, Hoefnagel portrayed an immense diversity of species, with as many as two dozen in a single scene. Shoals of minute yet precise fish are carefully fitted together in the space with blue water rippling around them.

There's a collection of flatfish – plaice, sole and flounder – showing the fine details of their asymmetrical faces with both eyes on the same side, some lying on their right and others on the left sides of their body, as different flatfish species do in the wild. A narrow pair of pipefish and a gurnard are joined by an assortment of blennies with their thick, pouting lips. Four pages are devoted to stingrays and skates, among them an electric ray with blue rings on its back and a cuckoo ray with a single black spot on each wing.

Several fish show that Hoefnagel had consulted other texts for details of what these animals looked like, including the books of Rondelet, Belon and Salviani. We can spot an eel coiled in spiral, a classic pose as Salviani and others had it.

Compared to those contemporary fish books and prints, Hoefnagel's gatherings of sea creatures served a very different purpose. Those other books aimed to identify and classify ocean species, and Hoefnagel drew on them for the information they contained but then elevated his paintings into a liminal space and blurred the lines between art and science. He was meticulous and accurate, while at the same time he created enchanting images that even those with little interest in the living world would surely find pleasing.

Some of the plates have biblical quotes, suggesting that Hoefnagel wished to spread an appreciation of the natural world in his audience. 'So is this great sea, which stretches wide its arms: there are creeping things without number: Creatures little and great' (Psalms 103:25, 26). 'Who shall be filled with beholding his glory?' (Ecclesiasticus 42:26). Hoefnagel's *Four Elements* were not intended to promote human domination and control over nature, but to close the gap between people and other animals, inviting viewers into a realm that they are part of and cohabit.

Several of Hoefnagel's paintings are notably ecological in their composition. One features a shoreline with a fabulous collection of invertebrates. Two distinct octopus species (with eight arms, suckers all the way along) and a cuttlefish (with eight arms and two tentacles, with suckers just at the tips) are floating in the open water. Sprouting up from the seabed are branching red

Joris Hoefnagel, Homelyn Ray and Four Other Rays or Skates, *plate* XXIX *from the volume* Animalia Aqvatilia et Cochiliata (Aqva), *c. 1575–80, watercolour.*

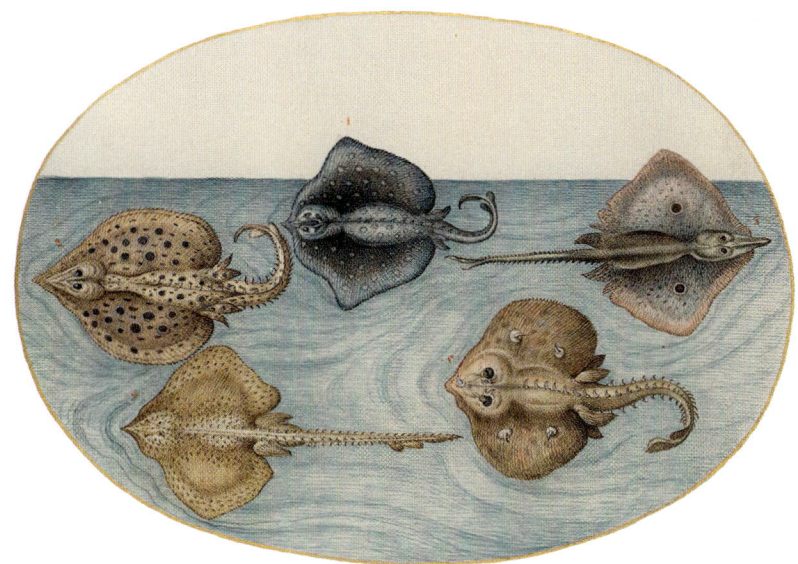

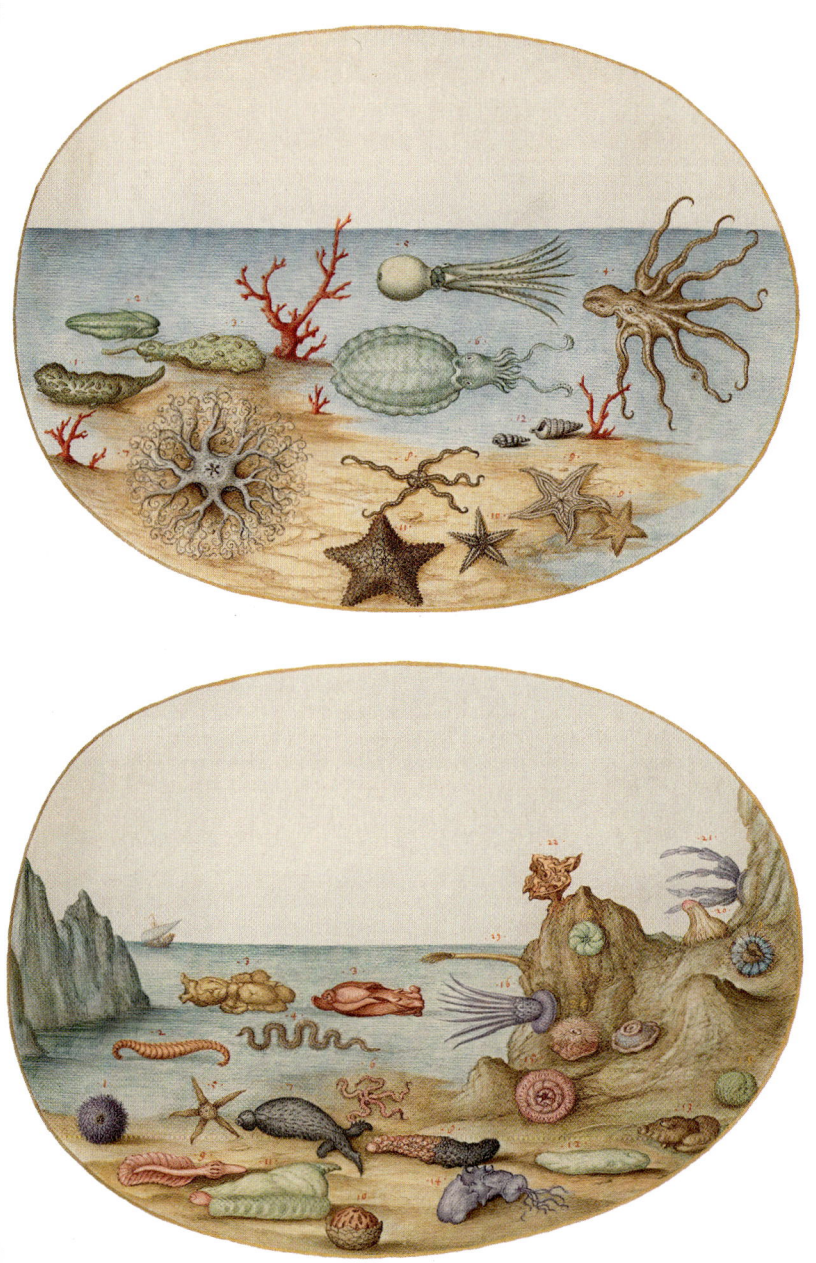

colonies of coral, and scattered about are seashells, starfish and various other less familiar echinoderms, such as the brittlestar, with five snakelike legs, and a splendid basket star, with a mass of twisting, curling arms. There are also sea cucumbers, which, despite their tiny size in the image, Hoefnagel painted not simply as amorphous, sausage-shaped blobs, but with accurate spiny textures to their skin and tentacles.

Sea cucumbers are even more splendid in another scene showing a rocky shore ecosystem. There's a pink warty sea cucumber, a green one that resembles a jelly mould, and a purple sea cucumber that has eviscerated itself and thrown its stomach through its mouth, a defence strategy seen in many sea cucumbers to smother attackers in sticky threads. A purple sea urchin sits by the waterline. A sea squirt lives up to its name and squirts a jet of water. Bristly worms wriggle through the sea. Two types of sea slug are swimming by. And fixed to the rocks are barnacles waving their legs, and several sea anemones have their tentacles opened out like flowers; others are withdrawn and look like blobs of goo, as they do when the tide goes out. All of that is packed into the 14 by 18-centimetre (5 by 7 in.) scene. This intricate and naturalistic assembly featuring so many creatures was way ahead of its time, foreshadowing a form of natural-history artwork that was still centuries off.

The Mimic Sea

More than two centuries after Hoefnagel was making his intricate miniature paintings, a technological advance took place that brought ocean life into view in a completely new way. In Europe

Opposite above: *Joris Hoefnagel*, Sea Cucumbers, Coral, Octopus, Starfish, Squid and Other Sea Creatures, *plate* LIII *from the series* 'Animalia Aqvatilia et Cochiliata (Aqva)', c. 1575–80, watercolour.
Opposite below: *Joris Hoefnagel*, Sea Cucumbers, Sea Urchins, Starfish, a Sea Nettle, a 'Sea Hare' and Other Marine Life, *plate* LIV *from the series* 'Animalia Aqvatilia et Cochiliata (Aqva)', c. 1575–80, watercolour.

in the first half of the nineteenth century, natural historians and collectors experimented with keeping creatures alive away from their ocean home. In the 1830s the French zoologist Jeanne Villepreux-Power began conducting pioneering studies of marine life at her home on the Mediterranean island of Sicily, where she invented the aquarium. She built glass and wooden boxes to study animals, including curious little octopuses called argonauts.

Living by the sea, Villepreux-Power had easy access to seawater to refresh her tanks and keep her animals healthy and alive. At a similar time, much further from the coast, another marine zoologist, Anna Thynne, began to experiment with keeping sea creatures in glass tanks at her home in London. She started by bringing in supplies of sea water from the coast, but soon decided that there must be a better way of keeping her tanks in good condition. She began experimenting with adding seaweed in the hope that it would enrich the water and ward off stagnation – and it worked. From the sea off Devon she gathered solitary coral polyps, which look like little stony flowers, and kept them alive in London for three years.

Thynne was the first person to make an ocean display that kept itself in balance. This was decades before the words 'ecology' and 'ecosystem' were coined, and yet Thynne grasped the principles behind these vital concepts of living things interacting with each other and their physical environment. She understood that the secret to a healthy living assemblage is to find the right mix of species that depend on one another.

Another British naturalist of the time, Philip Gosse, is credited with bringing ocean life and aquariums firmly into the public eye. Gosse spent much of his childhood on the coast in Dorset with his aunt, Susan Hill, who nurtured his love of nature and encouraged him to watch and draw what he found. As an adult, he spent time living overseas, including in Jamaica, where he collected specimens for shell collector Hugh Cuming. Returning to England, Gosse spent time in the West Country and in 1853 published *A Naturalist's Rambles on the Devonshire Coast*, inviting his readers to explore the wonders of the shoreline in his written descriptions and colour illustrations. In this book it is clear that Gosse saw no need to set a boundary between appreciation of the beauty of sea creatures and a scientific understanding of their lives. He wrote, 'That the increase of knowledge is in itself a pleasure to a healthy mind is surely true; but is there not in our hearts a chord that thrills in response to the beautiful, the joyous, the perfect, in Nature?'

In the same year, Gosse helped to establish Britain's first public display of sea life, at Regent's Park in London. Known as the London Fish House, the steel-and-glass construction housed wall-mounted and tabletop tanks that displayed all sorts of animals, including more than 4,000 specimens that Gosse had collected from the shores of Devon and brought to London by train. Members of the public flocked to the Fish House to gaze at these aquatic marvels, the likes of which most of them had never seen before.

Following the success of the London Fish House, in 1854 Gosse published *The Aquarium: An Unveiling of the Wonders of*

'The ancient wrasse', colour lithograph from Philip Henry Gosse, The Aquarium: An Unveiling of the Wonders of the Deep Sea, *2nd edn (1856).*

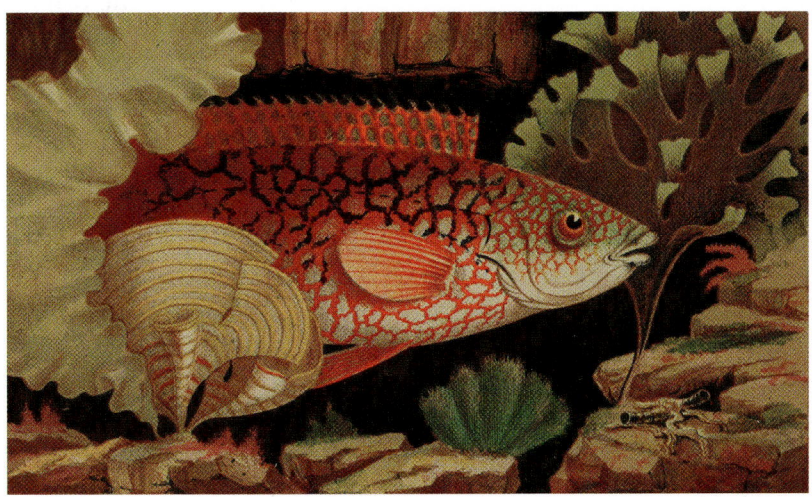

the Deep Sea (which was originally to be called *The Mimic Sea*), a guide to the theory and practice of keeping home aquariums. His was the first use in print of the word 'aquarium', a shortening of 'aquatic vivarium', a term people had been using until then for these water tanks filled with life. 'Aquarium' was, he wrote, 'neat, easily pronounced and easily remembered, significant and expressive'.

The Aquarium became a bestseller and was praised especially for its chromolithographic plates, prepared from Gosse's water-colour paintings. 'The volume ought to be on the table of every intelligent seaside visitor,' wrote a critic in *The Globe* in June 1854. The images show, again, Gosse's views of science and art. There is a mixture of detailed diagrams, such as a white sketch on black of a crab's mouthparts with 'A bristle magnified', together with natu-ralistic scenes that open a window for readers into living, under-water ecosystems: a ballan wrasse with honeycomb red patterns nestles among fronds of colourful seaweed; a painted topshell and white dorid nudibranch (a type of sea slug) carefully creep past the stinging tentacles of sea anemones. In the preface for the book, Gosse wrote that no one before him had attempted 'to represent marine animals, with their beauty of form and brilliance of colour, in their proper haunts, surrounded by sub-marine rocks and ele-gant sea-weeds, as these appear when transferred to an Aquarium'.

A reviewer in the *Literary Gazette* wrote in July 1854, 'Mr Gosse has himself dived into the bejewelled palaces which Old Neptune has so long kept reluctantly under lock and key, and we find their treasures set before us with a freshness and fidelity which afford welcome and instructive lessons to naturalists of all ages.' Gosse didn't actually plunge down to plunder Neptune's hoards but ventured offshore in a small boat and lowered down a dredging net to gather animals from the seabed.

Gosse spent months tinkering with aquarium tanks, hunt-ing for exactly the right balance of different species to keep the water clear and the specimens alive. Eventually he found a way to house a rich mix of animals and algae together in a 14-gallon tank: limpets, lobsters, scallops, sea urchins, swimming crabs, hermit crabs, spider crabs and many different types of seaweed. Once up and running, his aquariums let Gosse watch, study and paint eco-systems that until then had been hidden out of sight in the shallow seas, and seen only by a few brave divers who at around that time were beginning to lumber into the water wearing cumbersome diving helmets.

Later in life, Gosse became preoccupied with his religious views of the world and published *Omphalos: An Attempt to Untie the Geological Knot*, in which he argued that fossils were not evidence of evolution but rather a feature on Earth that God created to make the world appear older than it actually is. Nobody really bought into his ideas and very few bought his book. Most of the copies ended up being pulped and sold as waste paper. Still, though, Gosse is remembered today as a great popularizer of science who combined his talents as a communicator and artist to persuade the general public to engage with nature and science.

After *Omphalos* he produced several more scientific publications, including another book featuring his stunning artworks: *Actinologia Britannica: A History of the British Sea-Anemones and Corals*. Colour lithograph plates of his watercolours in *Actinologia* build on what Joris Hoefnagel started centuries earlier. While he didn't ring them in gold and his pictures weren't quite so small, Gosse created his own enclosed seascapes that envelop the viewer's gaze, showing us what it would be like to be under the sea, pressing our faces into these hidden, living worlds.

Around the ocean, there are more than a thousand species of sea anemone, relatives of corals and jellyfish (members of the phylum Cnidaria), all of them sedentary predators with tentacles covered in stings to grab and subdue small prey passing by. In tropical seas, many large species of anemone have fish living happily among their tentacles, as seen in the illustrations of the British scientist William Saville-Kent in an 1893 book describing his studies of the Great Barrier Reef in Australia. Just how anemonefish, also known as clownfish, avoid getting stung is a puzzle that scientists haven't fully solved; likely this involves the fish chemically disguising themselves as part of the sea anemone's body by smearing themselves in mucus from the base of the sea

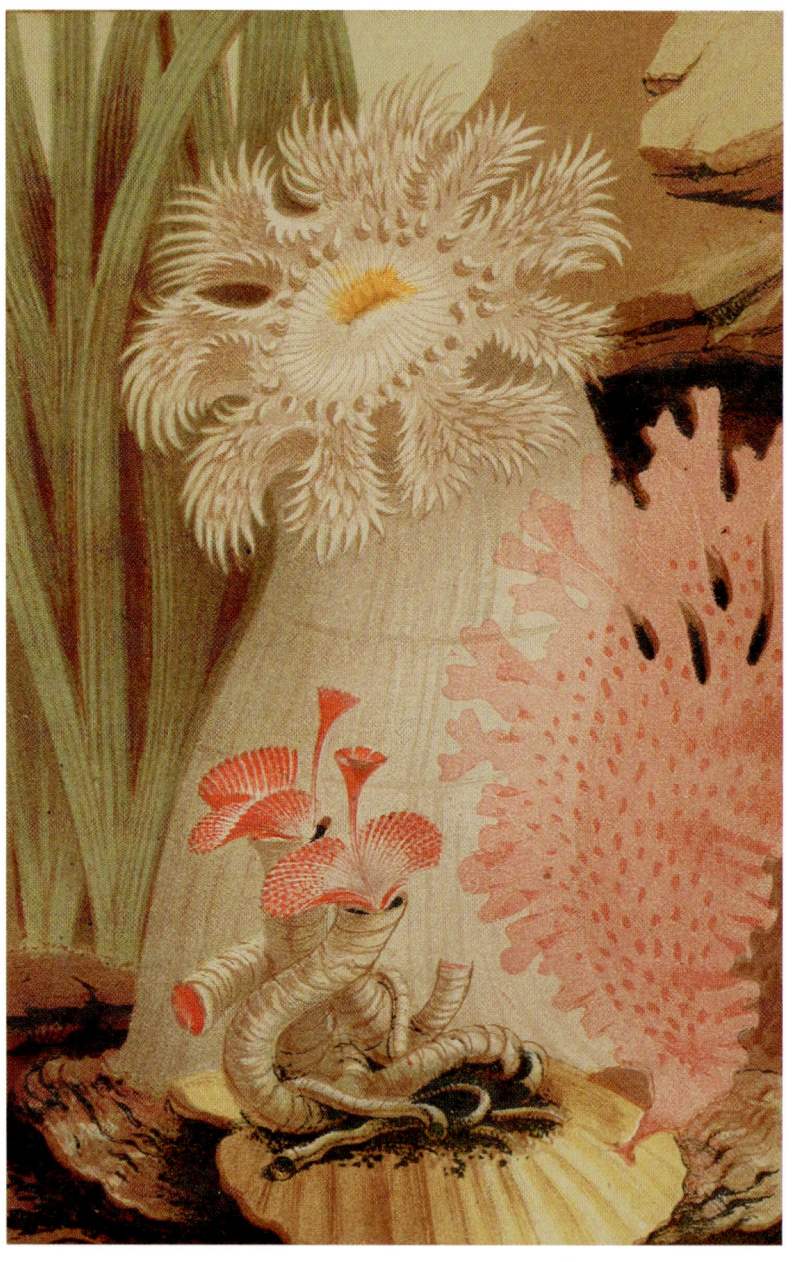

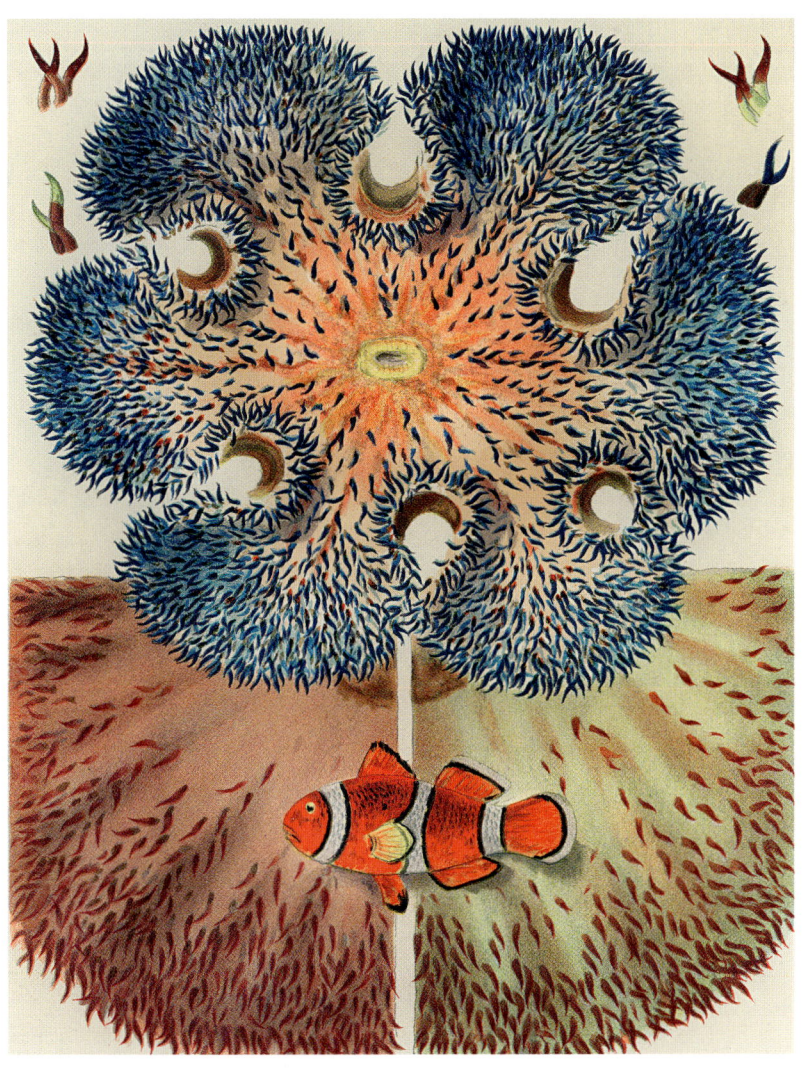

Opposite: *'The plumose anemone', colour lithograph from Philip Henry Gosse,* The Aquarium: An Unveiling of the Wonders of the Deep Sea, *2nd edn (1856).* Above: *'Barrier Reef Anemones', colour lithograph from William Saville-Kent,* The Great Barrier Reef of Australia *(1893).*

anemone. The most familiar sea anemones in Northern Europe are the shoreline species, such as beadlet anemones and strawberry anemones, both ruby red; many more live further down beneath the waves.

Gosse's sea anemones are pinpoint accurate, as useful as a clear photograph for identifying species that naturalists find in the wild. At the same time, Gosse's illustrations offer us something more. We can imagine what it would be like to poke a finger into those tangles of tentacles and feel a gentle suck from the stings that to us are harmless. We get a sense for how these species naturally fit in and around each other, building the layers of an ecosystem. Gosse has subtly and skilfully tidied things up for us, just a little – not so much that the arrangements of animals and seaweeds are too neat to be believable or too obviously sculpted for artistic effect. He makes us feel like we're really there. Gosse captured the shallow seas in a way that no one had quite managed before, turning them into realistic and stunning works of art.

Watching through Glass

The glass walls of aquariums offer artists the opportunity to observe living aquatic life for as long as they wish. Even though creatures are extracted from their realm and kept in a confined replica of their wide seas, given enough space and cared for by well-meaning aquarists it is possible to see how these species naturally move and behave, and to capture an essence of their lives that would be a great deal more challenging to observe in the wild.

In 1838 a zoo was founded in Amsterdam under the original name Natura Artis Magistra (meaning in Latin 'nature is the teacher of art') – these days it is generally just known as Artis. When the zoo's aquarium opened in 1882 many artists walked through its doors to watch, sketch and paint the aquatic animals and their realistic, although captive, surroundings.

Willem Witsen captured the thick-lipped pouts and despondent eyes of two Atlantic cod peering through a dark and weedy scene. In 1883 he made several paintings of wolffish (*Anarhichas lupus*), substantial denizens of the North Atlantic that can grow to more than a metre long and are named for their fang-like teeth, which they use to crush the hard shells of crabs, sea urchins, clams and whelks. Witsen's wolffish rest upright on the bottom of the aquarium, with bold blue-grey stripes across their eel-like bodies and delicate wrinkles of skin radiating from their eyes.

Four years later, August Legras produced an etching of an aquarium scene with cod and flatfish and a glum-looking wolffish, quite possibly the same individual that Witsen had studied (the species can live in the wild for twenty or thirty years, and treated well in captivity should last just as long). Legras spent several years studying the animals at Artis and was given a studio to work in. His contemporary, Gerrit Willem Dijsselhof, also paid regular visits to the aquarium. An archive of Dijsselhof's paintings, etchings and sketches shows his devotion to capturing the movements, shapes and light of these underwater scenes. He made dynamic

Willem Witsen, Wolffish at the Bottom of an Aquarium, *1883, watercolour.*

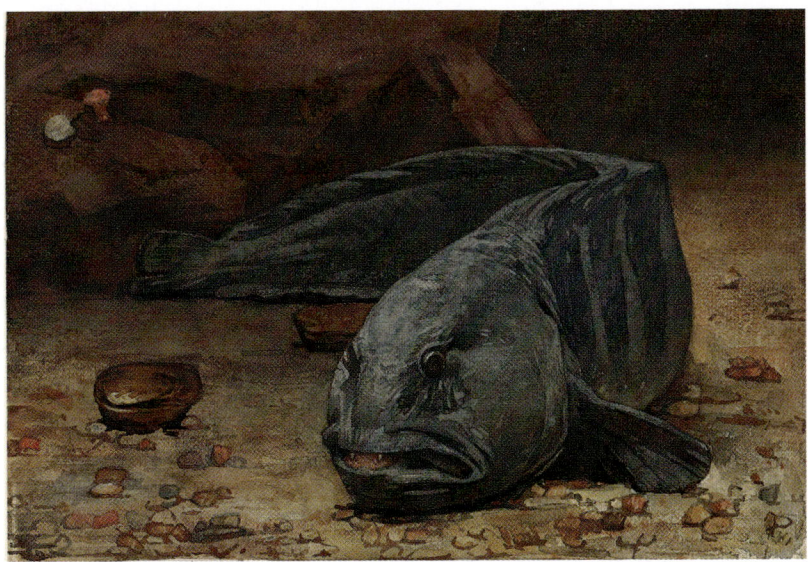

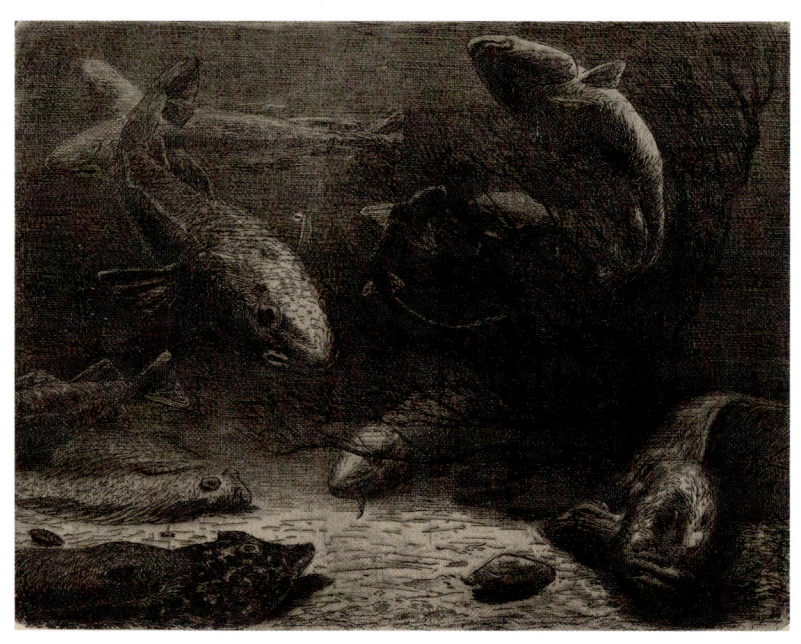

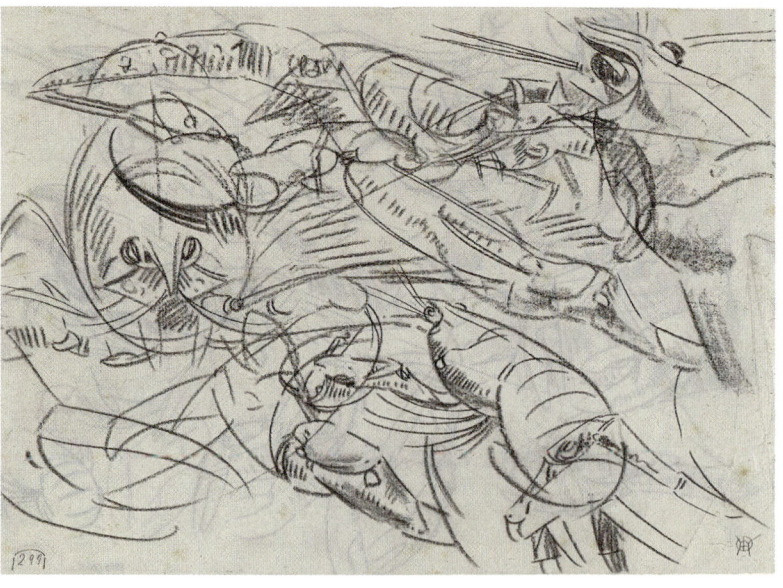

sketches outlining the flickering appendages of shrimps and showing how lobsters walk across the seabed with their tails unfurled,
and like to hunker into dark crevices, their snapping pincers and
long antennae facing outwards.

Dijsselhof noticed that well-fed aquarium inhabitants can
grow accustomed to each other and sit peacefully together, like
the catshark and the cobalt blue lobster, which in the wild are
more likely to keep their distance.

A chalk sketch shows that he understood a mannerism of
fish shoals that marine biologists have worked hard to grasp.
While the shoal itself seems to act as a single, coordinated being

Opposite above: *August Legras,* Aquarium, *1887, etching.*
Opposite below: *Gerrit Willem Dijsselhof,* Sketches of Shrimp,
n.d., chalk on paper. Below: *Gerrit Willem Dijsselhof,* Blue Lobster,
Dogfish and Flatfish in an Aquarium, *n.d., watercolour.*

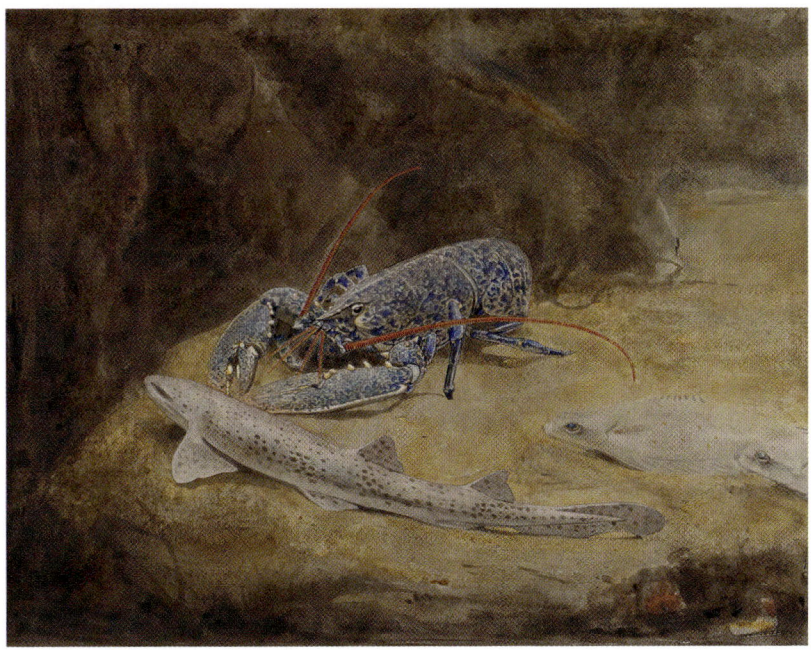

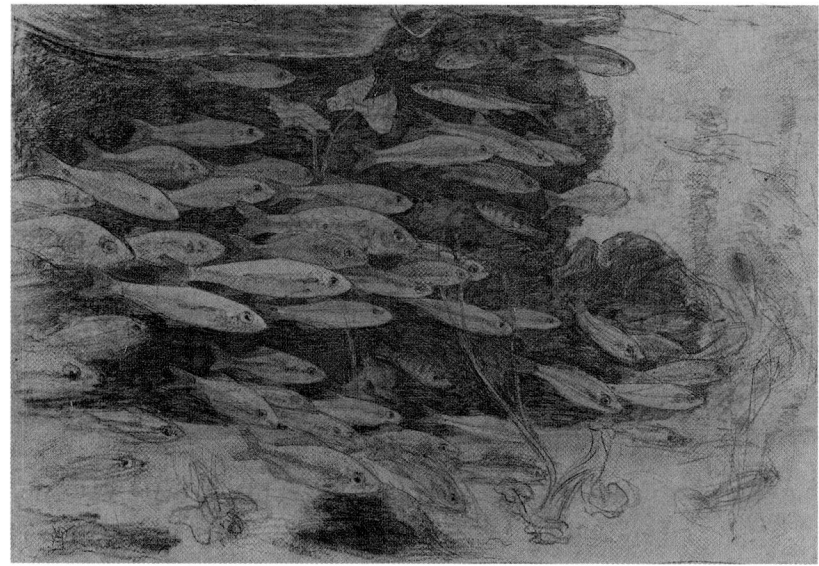

that shifts this way and that with clear intent and direction, it is nevertheless made up of individual animals doing their own thing and making their own decisions. Each of Dijsselhof's fish has its own expression and position in the shoal, frozen in a moment of perpetual motion.

Ocean Life Assembled

Philip Gosse's artworks, studies and aquarium guidebooks stirred great interest in ocean life among members of the British public who embraced aquarium-mania and ventured to the coasts to explore and collect. Aquatic pursuits became all the rage on the other side of the Atlantic too. In the mid-nineteenth century, North American public aquariums opened and books were published encouraging the same adventures and knowledge gathering as Gosse's works.

One of the earliest American illustrations of the shallow sea is *Ocean Life*, a gloriously colourful lithograph based on a watercolour painting by the French-born Philadelphia artist Christian Schussele. The image was created to accompany a fifty-page pamphlet published in 1859 by the physician and keen amateur naturalist James M. Sommerville. The image stands out for the immense diversity of species on display and the vivid sense it delivers of being underwater.

In the pamphlet's introduction, Sommerville describes his portrayal of the species depicted as in accordance with their 'natural size, and as they appear in their native element.' He does, however, admit that he assembled these species from localities widely separated – Florida, California, Mexico, Peru, Chile and Brazil – making this an imaginary, idealized ecosystem, intended to help people learn and study. He writes, 'The grouping may be considered, rather as an intellectual truth than as a literal view. – It might be termed a mental oasis of the sea.'

Opposite: *Gerrit Willem Dijsselhof,* Orfe Shoal in an Aquarium, *n.d., chalk on paper.* Below: *James M. Sommerville and Christian Schussele,* Ocean Life, c. *1859, watercolour.*

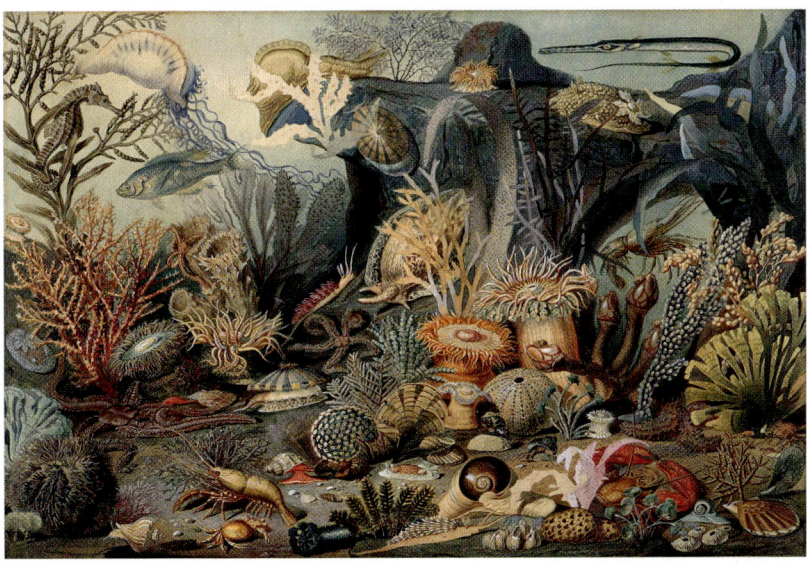

Faced with the busy scene of *Ocean Life*, it is difficult to decide where to look first. I get the exact same feeling when confronted with a flourishing marine ecosystem in real life, seen through the glass in my dive mask, except in those situations I have the added pressure of spotting living creatures before they swim away or scuttle off and hide. It is a lot easier when you can take your time and keep looking back at an unchanged scene in front of you. Sommerville and Schussele brilliantly created a scene that shows viewers what it's like to try and recognize species from afar.

Looking at *Ocean Life*, I start by picking out the bigger organisms that I know of. In the top right above a large rock floats the slender body of a cornetfish (*Fistularia tabacaria*), a distant relative of seahorses that has a similar long, tubular snout. Creeping across the rock below is a large, white dorid nudibranch (*Doris fontainii*), a type of sea slug with gills sticking up at the back like a feathery tail and a pair of ear-like appendages at the head end that act like a nose and sniff for chemical traces in the water. At the entrance to a small cave stand four thick stalks of gooseneck barnacles (*Pollicipes elegans*). Across to the left, a lined seahorse (*Hippocampus erectus*) wraps its tail around a clump of branching brown seaweed (*Stephanocystis osmundacea*). In the centre, clinging to a vertical rock face, is a keyhole limpet (*Fissurella costata*) with blue stripes radiating from the hole in the middle of the shell.

A rather misplaced Portuguese man-of-war (*Physalia physalis*) appears towards the top left corner of the image. A distant relative of jellyfish (a siphonophore), they normally float on the sea surface buoyed up by a pink gas-filled balloon, which in *Ocean Life* looks like a pale-coloured, crimped Cornish pasty. Sommerville and Schussele transplanted theirs underwater presumably for aesthetic reasons, a move that these jelly creatures can make by deflating their balloons so they can briefly sink and avoid airborne attacks, and yet this one is drifting below the waterline despite apparently being fully inflated. This Portuguese man-of-war is trailing its blue spiralling tentacles perilously close to a blue and yellow fish, the creole wrasse (*Clepticus parrae*). A brittlestar catches me by surprise – its five arms look far too big – but I check and see this is a tropical species I've not come across before. *Ophiocoma echinata* can be larger than my out-stretched hand, a giant compared to the little specimens I commonly find wriggling under rocks on the fringes of the eastern Atlantic.

Next in *Ocean Life* I turn my attention to spotting smaller species that require more careful consideration and closer scrutiny. I can spot at least two nudibranchs of the type with rows of feathery extensions all along their backs that serve multiple functions, including absorbing oxygen, digesting food and warding off intruders with stings that they get from their prey (often sea anemones and their kin), repurposing them as weapons for self-defence. There are distinct tufts of seaweed: in the centre of the scene is a clump of *Halimeda* that looks like little green coins fixed together in strips; bottom left are tiny, intricate green caps of *Acetabularia* that are made from single, giant cells and are known as mermaid's wineglasses.

In all, there are 75 species to find in *Ocean Life* and I confess I turn to Sommerville's list to help me spot many of them. A pleasing number of sea snails are prowling about the scene, not tucked up in their shells as they're often depicted, but showing their colourful foot and beady eyes: nutmeg snails, necklace snails, top shells, winkles and murex.

There are things I've never seen before. A sea pansy, or throne of stars as Sommerville calls them (*Renilla reniformis*), a rounded blue blob in the centre of the scene covered in little white flowers (and with a little crab hiding under it), is a type of soft coral. After storms they wash up on beaches in Florida and glow green when gently nudged, thanks to a mix of bioluminescent and fluorescent chemicals in their bodies.

By the time I have picked my way through the image and worked out what most things are, I'm convinced of the great value of *Ocean Life* in assisting the endeavours of sea-life spotters from the comfort of their armchairs. Then, I once again take in the whole scene, as I would if I were about to leave the water and return to land, and remind myself that this is more than a species key but

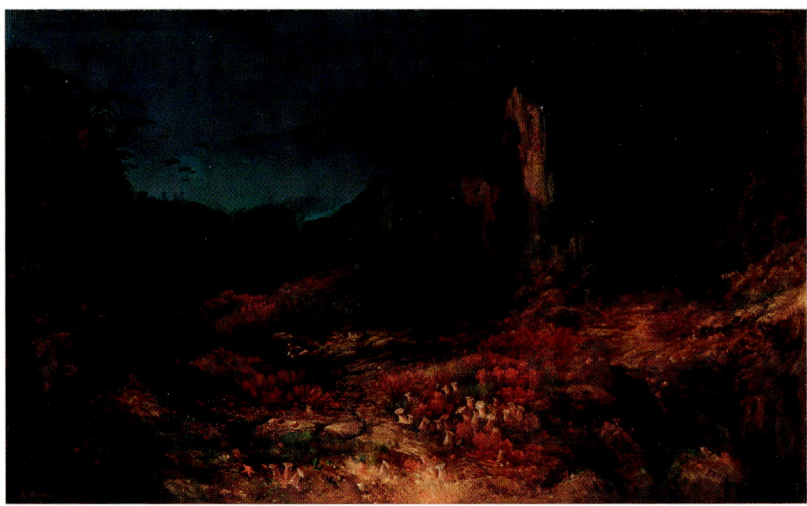

Edward Moran, The Valley in the Sea, *1862, oil on canvas.*

a beautiful arrangement of living forms and colours. They might not all be creatures that live together in the exact same part of the shallow seas, but that doesn't matter for the sake of this stunning image. Sommerville wasn't wrong when he wrote on the front page of his pamphlet, 'There's beauty in the deep.'

Three years after he published *Ocean Life* Sommerville bought (and perhaps commissioned) another painting by an American artist of an underwater scene with a rather different atmosphere, one that I think draws on the same idea of leading the viewer to ponder and appreciate the wonders of the ocean, although in a more subtle and surprising way.

Edward Moran's 1862 *The Valley in the Sea* looks at first glance like a terrestrial landscape scene, with a blooming flower meadow in the foreground, towering cliffs to the right and a darkening sky above, with clouds and birds in the distance. In fact those are not flowers or fungi, but sea anemones with sea snails and hermit crabs creeping among them, and the occasional starfish here and there. Those are not trees and plants but seaweed, and those are not birds but of course fish swimming past.

The Valley in the Sea feels to me like the beginning of a night dive, and simulates the sensation of taking to the water shortly after sunset. The idea of diving at night is instinctively and understandably frightening, but I find in reality it can be the most mesmerizing and calming time to be in the water. The quietness in the scene in the painting fits with the pause in activity between day and night shifts underwater. The diurnal species have taken themselves off somewhere sheltered to rest and sleep; the nocturnal species haven't yet woken up. For a while there is hush.

THREE

OPEN SEAS

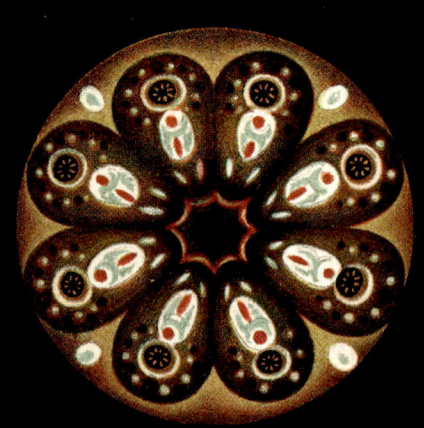

Before we embark on a journey towards more distant waters, I want to pause a while longer in parts of the shallower seas where artists have gone to great lengths to immerse themselves and create images *in situ*, as well as others who sink their installations into the ocean, bringing the viewers themselves beneath the waves.

From a young age, Eugen von Ransonnet-Villez had two great loves: art and the sea. He was born in 1838 into an aristocratic family in Vienna who sent him to study at the Academy of Fine Arts when he was only twelve years old. Frequent visits to his aunt on the French Mediterranean coast nurtured his interest in sea life. Ransonnet-Villez's family persuaded him to train as a lawyer and he took employment as an official at the Imperial Ministry of Foreign Affairs, which gave him the chance to travel extensively overseas and pursue his interests in the ocean.

In 1862, during a trip to El Tor on Egypt's Sinai peninsula, Ransonnet-Villez visited the Red Sea and saw his first tropical coral reef. On his return to Europe, he published a book about his travels that included several colour lithograph plates based on his watercolours of coral reefs. He had stayed above the waterline and from a boat peered down through the clear waters of the Red Sea. These artworks capture a similar spirit of detail and realism as James Sommerville and Christian Schussele's *Ocean Life*, and continue the same ambitions as Philip Gosse to both dazzle and educate people about otherwise inaccessible life in the ocean. Ransonnet-Villez wrote in 1863 in the journal of the Royal Zoological-Botanic Society of Vienna, 'The illustrations presented here show the seafloor like on the inside of a fish tank, in the same perspective arrangement as it is used in landscapes.'

Ransonnet-Villez evidently had greater ambitions for his sealife paintings, which he set about fulfilling on a trip two years later to Sri Lanka. In the southern coastal city of Galle, he designed and had built a one-person diving bell. Made from a sheet of iron and weighed down by six cannonballs slung in cloth bags, the angular structure had just enough room inside for Ransonnet-Villez to swim down and pop his head and torso into the bubble of air. Fresh air was pumped down by hand from a boat at the surface. When he wanted to move, he planted his feet on the seabed, picked up the whole device and marched forwards a few paces. Then, sitting on a strap and dangling his feet, the artist stayed underwater for as long as three hours at a time, gazing through a small circular glass window and drawing the corals and fish outside. He brought down pencils and paper, which he waterproofed with varnish.

Back on land, he went over his sketches with oil paints and recreated a view of the underwater world that art viewers had never seen before. Undoubtedly, Ransonnet-Villez cared about accuracy and detail, and his paintings depict recognizable species. He combined this scientific precision with keen-eyed observations of living, moving animals, as other artists were doing in the confines of aquariums, only his animals were wild and free. He also expertly depicted the particular way that light behaves in tropical seas. The water is clear but has a certain kind of haziness

'Submarine rocks with green corals', colour lithograph from Eugen von Ransonnet-Villez, Sketches of the Inhabitants, Animal Life and Vegetation in the Lowlands and High Mountains of Ceylon *(1867).*

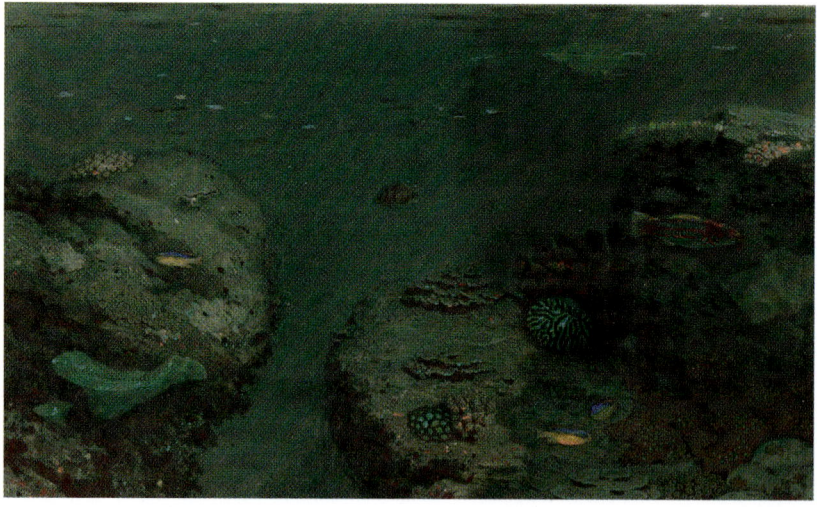

and shifts towards blue reaching into the distance, in a way that's very different from looking through air. And the sea surface overhead is a rippling mirror, reflecting what lies below, somewhat like looking down on the surface of a pond, but not quite. Ransonnet-Villez's paintings bring his viewers into the water and has them sit next to him inside that cramped diving bell, looking out onto a real coral reef.

Several decades later, another artist took his paints to the deep. A cartoon from a 1921 edition of the French weekly news magazine *Le Pèlerin* shows Zarh Pritchard dressed in a deep-sea diving suit, a round-windowed helmet on his head and air pipe leading to the surface. He stands feet apart on the seabed with his easel and palette, surrounded by assorted puzzled-looking fish peering at his painting. This is probably how Pritchard would have wanted to be remembered. In his lifetime (1866–1956) he made a name for himself as the world's leading and at that point perhaps the only underwater painter.

Born in India to British and Irish parents, Pritchard was schooled in Scotland, where he got his first glimpses of the world beneath the waves. He jumped into in the chilly waters of the Firth of Forth, swimming underwater with his eyes open, then later making for himself a basic pair of diving goggles. After each plunge, Pritchard hurried back to land to sketch pictures of what he had seen. He studied at the Edinburgh College of Art, designed sea-themed costumes for the French actress Sarah Bernhardt and in the early 1900s moved to California. On travels to Tahiti in 1904 Pritchard began making art underwater. Initially, he would swim down, holding his breath, and sketch with crayons on oiled paper taped to a sheet of glass. Soon he met Narii Salmon, brother of the former queen of Tahiti. Together they explored the coral reefs and Pritchard borrowed from Salmon the only diving suit in the islands. This allowed him to descend to the sea floor, 10 metres (33 ft) down, for half an hour at a time, carrying with him his oil paints and easel.

Similar to Ransonnet-Villez's diving-bell artworks, Pritchard worked on his underwater sketches back on dry land. The paintings

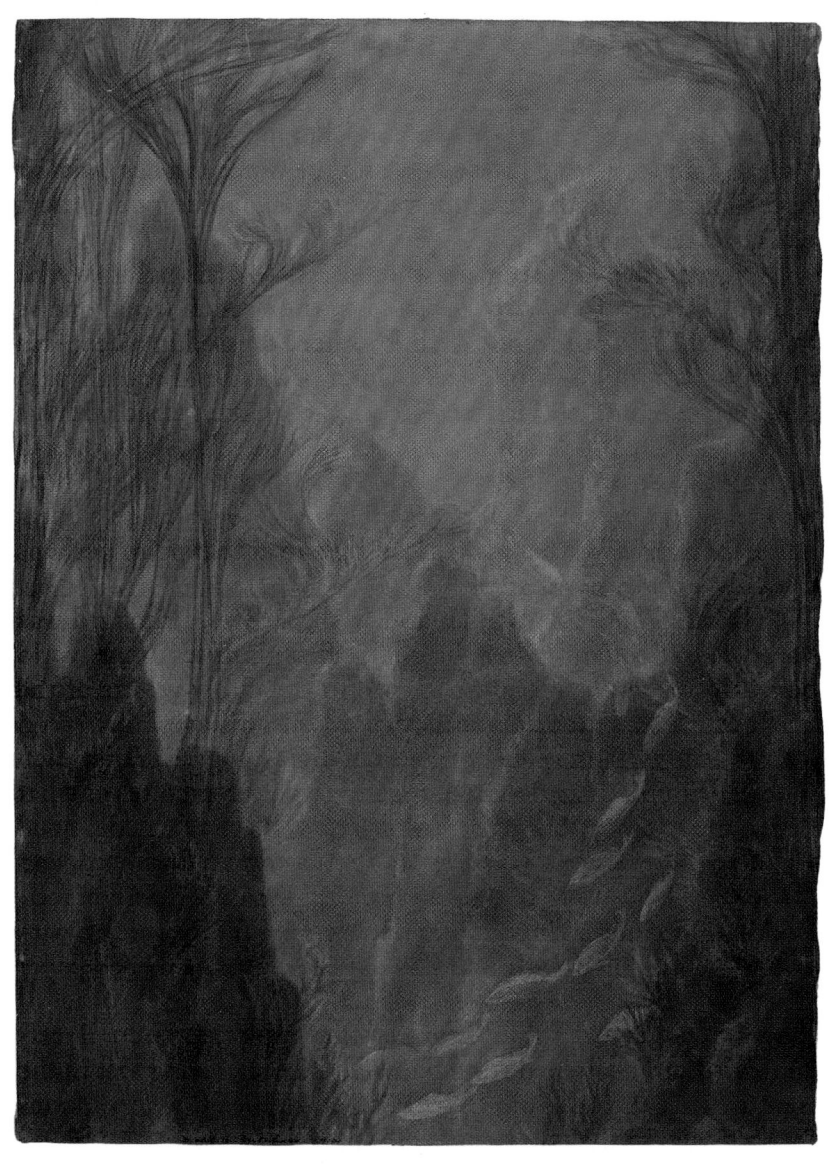

Zarh H. Pritchard, Bream in Twenty-Five Feet of Water
Off the West Coast of Scotland, *1910, pastel.*

he produced were popular among natural-history museums in the United States and Europe. Many paintings were bought for the Musée Océanographique in Monaco.

Pritchard's paintings are gentle and pastel-coloured, with more than a hint of a fairy-tale dreamland about them, or the atmosphere of underwater shell grottos. The images I like best are not those in which Pritchard painted real fish species – a small shoal of red squirrelfish, or a pair of white, black and yellow striped fish called Moorish idols. Those images feel too washed-out and underwhelming; they lack the radiant quality that coral reef fish emit when seen up close underwater. I prefer Pritchard's underwater scenes that remind me of a hazy tropical dive when the water is misted with particles stirred from a recent storm, and, looking into the distance, the colours of a reef are pastel and muted.

Contemporary British sculptor Jason deCaires Taylor specializes in producing large installations from pH-neutral marine-grade cement, which he fixes on the seabed, where they transform into living artworks in collaboration with sea life itself. Seaweed spores and larvae from corals and sponges settle onto the structures, encrusting and ornamenting his works. DeCaires Taylor designs nooks and hideaways deep inside the works to encourage fish and lobsters to move in and occupy his creations.

His sculptures commonly feature casts of people who live along the coasts neighbouring his underwater museums. Off the coast of Cannes, France, six giant mask-like faces sit on sandy patches of seabed among swirling meadows of *Posidonia* seagrass. The works installed in 2021 are portraits of local residents, including an eighty-year-old fisherman and a nine-year-old child. In the waters of the Cancún Marine Park in Mexico is the Museo Subacuático de Arte, a collection created in 2009 of more than five hundred sculptures. Among the works, *The Silent Evolution* features statues of the local fishing community, Puerto Morelos, who now stand together on the seabed, defending their ocean and over time turning into ocean life themselves, their faces morphing into coral reefs.

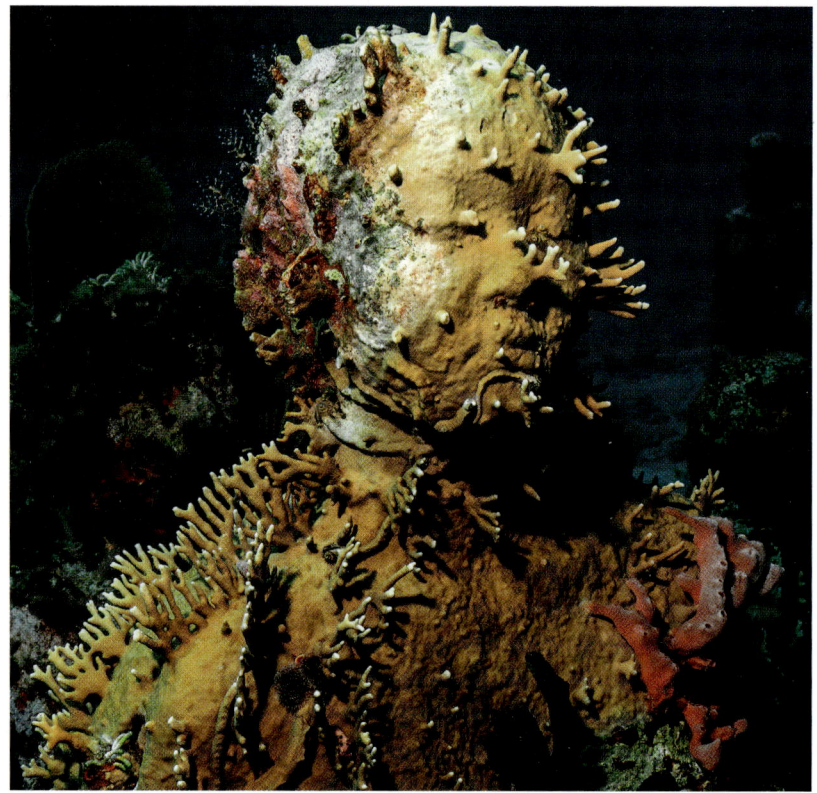

Underwater sculpture of local resident turning into a living reef by Jason deCaires Taylor, part of the Silent Evolution *installation, 2012, photograph dated 2022, Museo Subacuático de Arte (*MUSA*), Cancún.*

Visitors swim through these underwater galleries, either snorkelling or scuba diving. In a practical sense, deCaires Taylor intends his works to relieve pressure on busy parts of the seas, where tourists flock to swim and dive and unintentionally damage delicate sea life. His works also persuade people to reflect on the connections between humans and ocean life. Certainly there's something striking and stimulating in encountering man-made artworks on the seabed. Objects always look very different underwater, because of the way light behaves; things appear closer and bigger than expected, and the light is always moving and shifting as it filters down through the waves. And it's intriguing to see sea creatures interacting with those structures and making them part of their world.

Human faces have also appeared on the seabed off the coast of Tuscany, Italy, these ones with a particular purpose in mind. Incensed by the industrial trawlers operating illegally close to shore and ripping apart the delicate seagrass meadows, local fisherman Paolo Fanciulli came up with an artistic solution. He set up a foundation and commissioned artists to carve dozens of sculptures from marble, and in 2015 began sinking them around the seagrass meadows. The large blocks prevent the trawlers from operating in these areas because their nets snag on the structures and break. Within a few years, damaged areas showed signs of recovery and regrowth of the seagrasses that lock up huge amounts of carbon from the atmosphere and nurture a critical habitat for many other underwater species.

Over the Horizon

Heading offshore, we reach the regions of the ocean that are by far the most substantial but the least visited by people. Beyond the edges of continental shelves the sea floor falls away towards the abyss, and lying above it is the vast blue blanket of the open seas. More than 200 nautical miles from shore lie the waters technically known as the high seas, which no nation can lay claim to, although some try, and which account for two-thirds of the ocean's surface area and cover almost half of the entire planet.

To get there, we could follow the trail of animals that journey from coasts and inland waterways out to the distant high seas. Eels of various species (*Anguilla* spp.) have been known of for centuries, trapped in rivers, eaten in huge numbers, and portrayed in heraldry and artworks with their sinuous bodies and small pectoral fins that stick out like ears. People have long puzzled over where eels come from and where they go after they set off downriver and swim for the sea. It was only in 2022 that scientists tracked European eels for the first time, using tiny satellite transmitters, migrating thousands of miles across the Atlantic to the warm, clear waters of the Sargasso Sea, to the east of the island of Bermuda. The Sargasso is the only sea that has no land boundaries and instead is circled in a loop of currents that sweep together a floating forest of sargassum seaweed, buoyed up by round gas-filled bladders that look like berries. Eels come all the way to the Sargasso Sea to spawn. Females and males release their eggs and sperm, then they die, leaving their fertilized eggs to hatch into larvae, which look like tiny transparent leaves with heads attached. Subsequently, the newborns have an immense journey ahead of them – the larvae drift on ocean currents all the way back to Europe.

We could also reach the Sargasso Sea by following turtle hatchlings. After scrambling out of their sandy nests on the warm beaches of Florida, Mexico and islands of the Caribbean, the palm-sized hatchlings hurry for the waterline and, using an in-built magnetic compass, head east to seek shelter and food in the floating forests. No artists have made that journey, but plenty have seen turtles closer to shore, before adults and young set off on great, ocean-spanning migrations.

In the 1720s the English naturalist Mark Catesby travelled around North America, gathering observations and specimens that formed the basis of a work that is generally credited with being the first publication describing and illustrating the flora and fauna of the region. He travelled along the east coast as far north as the Appalachian Mountains, and later went to the islands of the Bahamas, for a time taking with him an enslaved boy to assist his work. Returning to Britain, Catesby spent twenty years producing *The Natural History of Carolina, Florida and the Bahama Islands*, published in multiple volumes paid for by subscribers and illustrated with his own copperplate etchings, most of which feature a pairing of an animal and a plant species, drawn in richly coloured detail and elegantly arranged together on the page. For the most part, these were not intended to be realistic, ecological

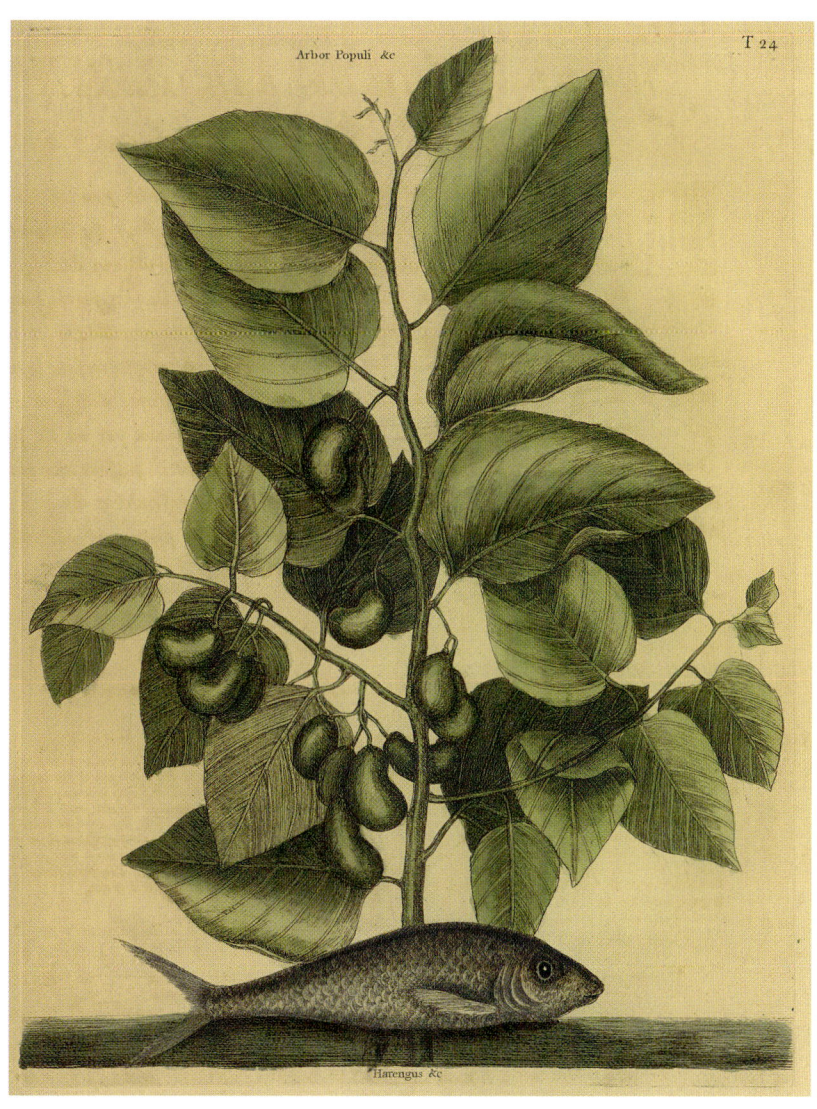

*'Pilchard (*Argentina carolina*)', hand-coloured engraving from Mark Catesby,* The Natural History of Carolina, Florida and the Bahama Islands, *vol. II (1743).*

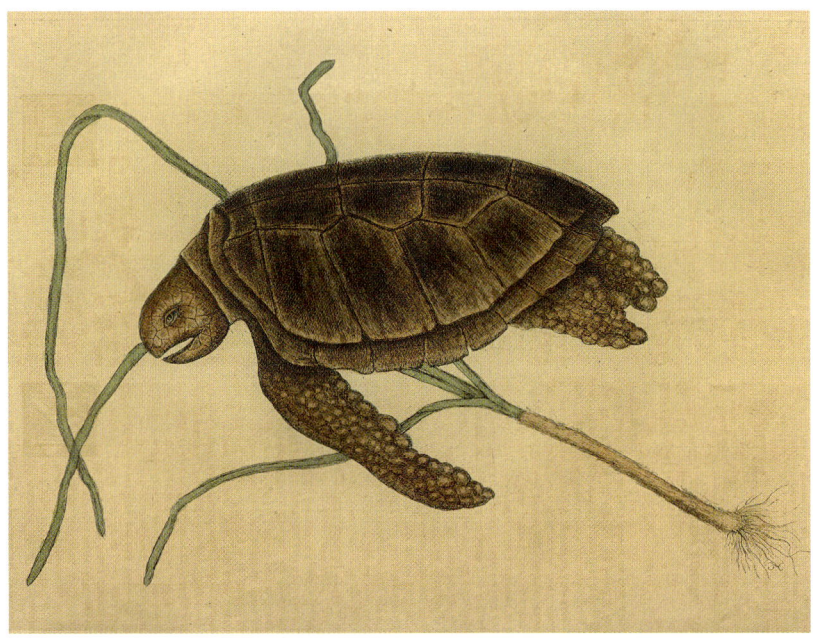

*'Green turtle (*Testudo marina viridis*)', hand-coloured engraving from Mark Catesby,* The Natural History of Carolina, Florida and the Bahama Islands, *vol. II (1743).*

depictions. Catesby sometimes etched fish species surrounded by terrestrial plants, such as a pufferfish (*Lagocephalus lagocephalus*) surrounded by branches of sassafras trees, and a pilchard with a species of shrub sprouting behind it. However, one plate shows a rather weary-looking green turtle (*Chelonia mydas*) and a sprig of seagrass, known as turtlegrass (*Thalassia testudinum*) because it forms a major part of their diet. He also drew a hawksbill turtle (*Eretmochelys imbricata*) on the beach next to a nest of eggs, and a loggerhead turtle (*Caretta caretta*) swimming in the shallows, perhaps a female arriving to nest.

Among Catesby's illustrations of North American birds, there are seabirds that would lead us offshore into the open seas. In one plate a brown noddy (*Anous stolidus*) stands on a rock, looking out to sea. This is a species of tern, unusual for its dark body and white-capped head (most terns are the opposite colouration), and they nest on tropical islands. Catesby would have come across many of them in the Bahamas and perhaps on islands off Florida. While their migratory behaviour is not well understood, brown noddies certainly spend a lot of time far out to sea, all around the tropics.

A century after Catesby, the American artist and naturalist John James Audubon produced another seminal illustrated natural-history book based on the wildlife of North America. Among the colour paintings in *The Birds of America*, Audubon included more than 170 seabirds. He intended each bird to be painted life-size and so he used the biggest available paper, more than 2 feet by 3 (roughly 91 × 66 cm), known as double elephant folio. The book, published between 1827 and 1838, is considered one of the most valuable natural-history books of all time and one of the most stunning artworks of birds ever made. Audubon's legacy as a co-founder of Western ornithology and figurehead for the modern conservation movement is coming under new scrutiny as more people are reckoning with his racist writings and involvement in the slave trade. His views of Black and Indigenous people were deeply troubling and he supported his work by buying and selling enslaved people.

Audubon crossed the Atlantic Ocean twelve times and during many of those voyages he observed seabirds in remote open seas – and shot them, as was common practice among naturalists at the time, so they could get a closer look at the animals. Storm petrels seem to have been his particular favourite. He painted three species, the European, Wilson's and Leach's storm petrels (*Hydrobates pelagicus*, *Oceanites oceanicus* and *Hydrobates leucorhous*). These are the smallest of all the seabirds, roughly the body length and one-ounce weight of a sparrow. They spend most of their lives at sea, hunting for small fish and squid, and only come to rocky islands and a few isolated headlands to breed and raise their young.

Audubon's painting of a pair of Leach's petrels (known then as fork-tailed petrels) shows them being buffeted by winds above stormy waves. He wrote careful notes about the species' breeding range, migrations and abundance, based on his own observations and information gathered from sailors and fishers he met. For scientists today, these are rare and largely accurate historical records from a time when there were few data on these birds that live most of their lives away from humans. Audubon's views of storm petrels also offer an interesting cultural insight on birds that in the past, and still in the present, many are fearful of. There's a long history of people giving petrels hateful names like devil birds and water witches, and believing that they are responsible for stirring storms. Audubon was undoubtedly cruel to many people in his writings, but he was clearly fond of these little birds. He fed them on the decks of ships, recognized their vulnerability in high winds and appreciated them as early warnings of bad weather on the way.

As well as travellers aiming to get to the other side of the ocean, another group of people who spent a lot of time on the open

*'Stormy Petrel (*Thalassidroma wilsonii*)', drawn from nature by John James Audubon, 1835.*

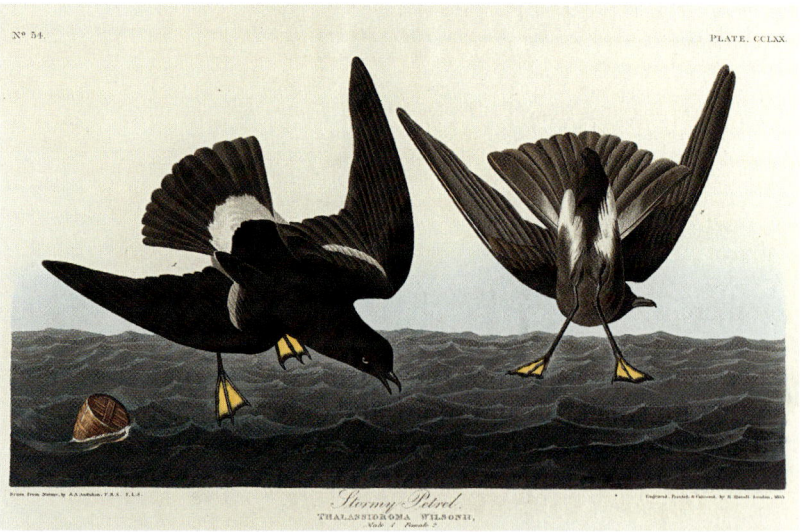

seas were commercial whalers from Europe and North America. During the height of the whale hunt, in the eighteenth and nineteenth centuries, people working in the whaling fleets lived at sea for years at a time. In between frenetic activity of hunting, then flensing and processing whale carcasses, there were long, boring interludes which creative members of the crew filled by making artworks from by-products of the whales they were catching. Scrimshaw is the art of carving designs into the teeth or bones of marine mammals, sometimes the stiff baleen plates from inside the mouths of filter-feeding great whales like humpbacks and blue whales. Scrimshanders, as the craftsmen are known, polished the surface of bones and teeth using shark skin as sandpaper, then etched designs using a sharp needle and rubbed in ink or soot to make them show up.

The earliest examples from European whalers came from Arctic voyages of the seventeenth century, and the practice carried on into the twentieth century when whales were still being hunted in the seas around Antarctica. Most scrimshaw was made with the stocky, pointed teeth of sperm whales on the long whaling cruises between the 1830s and 1870s. Popular subjects were scenes from their voyages, as well as fashionable men and women, probably copied from books and magazines. Ships often featured either singly sailing across the waves or in the midst of a whale hunt. And some scrimshaw focused on the whales themselves.

There are thirteen species of great whales roaming the open seas today, among them the biggest animals ever known to have existed, blue whales. All the great whales are typically longer than 10 metres (32 ft) from head to tail and weigh 10 or even as much as 200 tonnes. Most species were the target at one time or another of commercial whaling operations, their bodies the source of valuable and culturally significant commodities. Their oil lit the streets of Europe and America, and strips of baleen plates shaped the exaggerated hourglass figures of corseted women. Unsurprisingly, many paintings and prints by European, American and Japanese artists depict whaling scenes, which are often shown as fearsome battles.

It is harder to find images of living whales disconnected from hunting activities, swimming freely through their realm of the open seas. They feature in natural-history illustrations and other artworks, although in older works much of their appearance would likely have been based on hunted animals and beach strandings. Until advanced diving equipment and underwater cameras came along, people would only have seen parts of living whales, their heads and tails reaching briefly above the waterline. This is probably why, even until relatively recently, depictions of whales often didn't look quite right. For instance, in the 1920 book *British Mammals*, Archibald Thorburn's colour plate shows rather portly bowhead and humpback whales (*Balaena mysticetus* and *Megaptera novaeangliae*).

Today it is difficult to imagine that only a few decades ago the ocean's great whales were widely regarded as little more than an industrial resource. Well into the second half of the twentieth century, whale oils and fats were routinely added to everyday products in Europe and North America. Car transmission fluids, the glue on stamps and envelopes, lipstick, margarine and children's colouring crayons could all contain residues of the biggest animals ever to exist.

As more people woke up to the cruelty of killing whales, the beauty of their underwater songs and the drastic declines in wild populations, Greenpeace's Save the Whales campaign kicked into gear in the 1970s and began what was ultimately a successful push-back against the whaling industry. In 1986 a moratorium made commercial whaling illegal. Since that time, the view of whales in the public eye has, for the most part, radically shifted, and these animals are now generally considered to be precious wildlife that are worthy of strict protection.

Underwater film-makers and photographers have brought images of these living animals into view, and it follows that many artists have also embraced the beauty and forms of whales. Among the many pods-worth of whales depicted in contemporary artworks, one that stands out to me for its impact and beauty is a mural painted by French artist Lily Mixe on a wall in London's

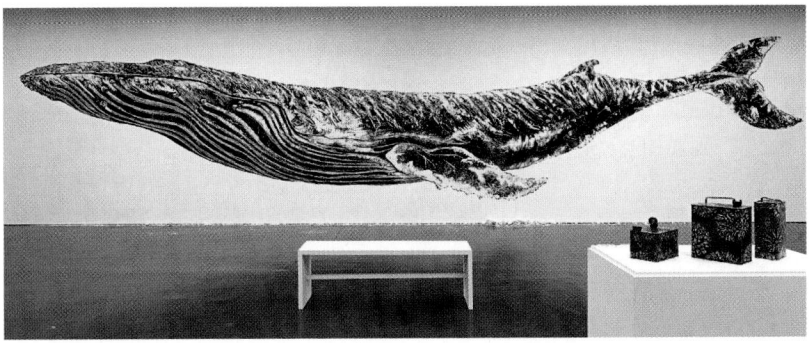

Lily Mixe, Life Size Humpback Whale, *mural, black masonry paint and chalk, part of the 'Butterfly Effect' exhibition, 2022, Saatchi Gallery, London.*

Saatchi Gallery in her 2022 'Butterfly Effect' exhibition. She created a life-size, 15-metre-long (50 ft) image of a humpback whale, using chalk from the coastline of her home base of Margate in Kent. This chalk is made from countless billions of fossilized shells of microscopic phytoplankton that were alive during the Cretaceous Period, around 80 million years ago. These particular plankton existed before whales, which evolved from land-based mammalian ancestors only within the last 50 million years. In the modern-day open ocean, phytoplankton are the basis of food webs. They harness energy from the Sun and create food for animals, including the small fish and shrimp that are eaten by filter-feeding baleen whales, such as humpbacks. Mixe's temporary artwork allowed viewers to stand on dry land and contemplate the whole of a whale, as well as the connections from these giants to the minute creatures that support the living systems on which they depend.

Creatures of the Furthest Reaches

As well as the great whales, there are various other species that can usually be seen only out in the open seas, far from shore, and yet they still make their way into many artworks. The idea of a flying fish seems rather improbable: fish that choose to leave their aquatic realm and have evolved wings that allow them to soar through the air (they belong to the family Exocoetidae). Indeed, the first time I watched as a flying fish leapt into view and skittered across the waves beside the boat I was on, I felt like I was gazing at a fairy tale. I've seen many since that first one and still each time it's a thrill to witness their brave transition from water to dry air.

Many natural-history artists have depicted these beautiful fish, often in flight above the sea. Flying fish perform their incredible feats by first generating speed underwater, beating their tails at up to seventy times per second, and aiming upwards to pierce the waterline. Then their enlarged pectoral fins open out – in some species accompanied by a pair of large pelvic fins, like a biplane – and they whiz along for several hundred metres. The question of whether flying fish are truly flying, and actively flapping their 'wings', kept biologists puzzled for some time – because these fish are so fast it is difficult to see what's going on as they zip through the air. High-speed cameras have shown, however, that they hold their fins still and are gliding. Flying fish also lack powerful flight muscles that birds need to flap their wings fast enough. Nevertheless, flying fish do have tricks to help them stay airborne for longer. Some artworks show the longer, lower lobe of the fish's tail fin (or caudal fin) dipping in the water. Swishing their tail gives them some added propulsion and helps them fly for longer. Generally, their flights last less than a minute, but that's long enough to escape underwater predators, especially with the added trick of changing direction mid-flight. Often flying fish veer off in a different direction from the one they set off in, presumably to throw off their pursuers.

Many species of pelagic shark are rarely seen in shallow and coastal waters, but spend their lives roaming the open seas. These long-distance endurance swimmers often make their way across entire ocean basins, travelling thousands of miles in a few months. Mariners see these animals during their own oceanic crossings and fishermen catch them when working out in the open seas.

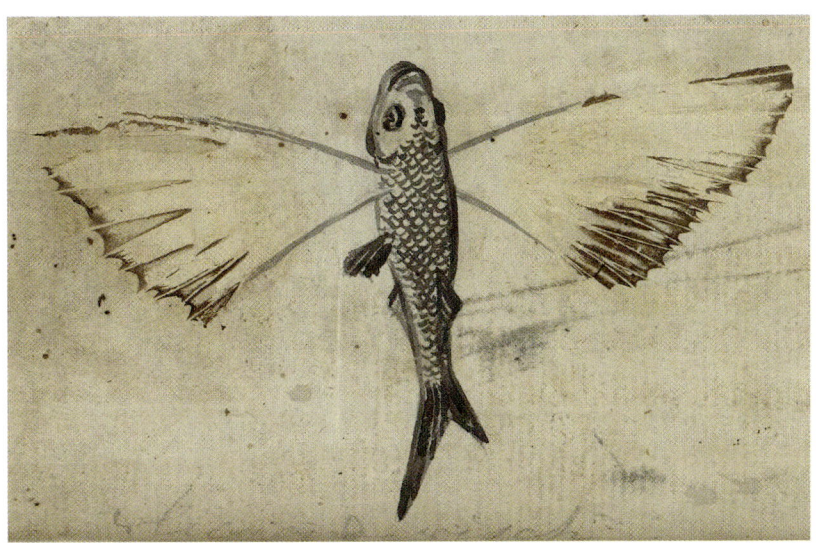

Jan Brandes, Sketch of Flying Fish, *1785–7, pencil and ink.*

This is how Cornish natural historian Jonathan Couch got his hands on specimens to study and illustrate in his four-volume book from the 1860s, *A History of the Fishes of the British Islands.* Several colour illustrations give elegant depictions of pelagic sharks, such as the sleek form of a blue shark (*Prionace glauca*), which today is one of the most abundant sharks in the open seas, and a thresher shark (*Alopias vulpinus*), a species with great long upper lobes to their tails that they snap over their head like a whip to stun shoals of fish.

The earliest known illustration of another pelagic shark species comes from the 1820s, and a French colonial expedition around the world on the sailing corvette *La Coquille* (The Shell). The colour engraving of an oceanic whitetip shark (*Carcharhinus longimanus*, then called *Squale maou*) clearly shows the elongated dorsal and pectoral fins, with dappled white patterns at their tips.

Until recent times, oceanic whitetips were considered to be one of the most abundant large animals on Earth. They vary in size, but can be as long as 3 or even 4 metres (10 or 13 ft) from nose to

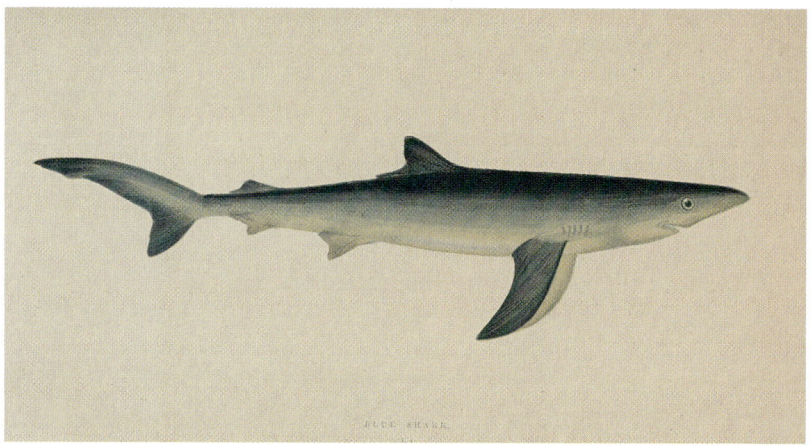

tail. Mariners often encountered these sharks because they have the habit of following ships and investigating any objects floating in the blue waters of the open seas; sharks gained a reputation, likely undeserved, for attacking and eating shipwreck victims. Oceanic whitetip sharks, which once roamed in great numbers through all the ocean, except the icy polar seas, have been driven towards extinction by rampant industrial fishing, in particular long-lining. A single fishing vessel can catch more than fifty of these big sharks in one go. Nowadays, they're much harder to find than they were a few decades ago.

Stories have been told for centuries about fish that hang out with the sharks of the open seas. Remoras, also known as shark suckers (from the family Echeneidae), are the origin of legends about creatures that latch onto the hulls of ships and slow them down. Roman writer Pliny the Elder blamed these ship holders for Antony and Cleopatra's defeat in a battle at sea that led to the end of the Roman Republic in 31 BC. At most a metre long, even a huge shoal of remoras won't be strong enough to halt a fleet of battleships; however, they do hitch a ride on large fish and dolphins, adding to their drag in the water. By doing so, remoras save energy on swimming, can steal bits of food from their ride and when they're going fast enough they can just open their mouths to

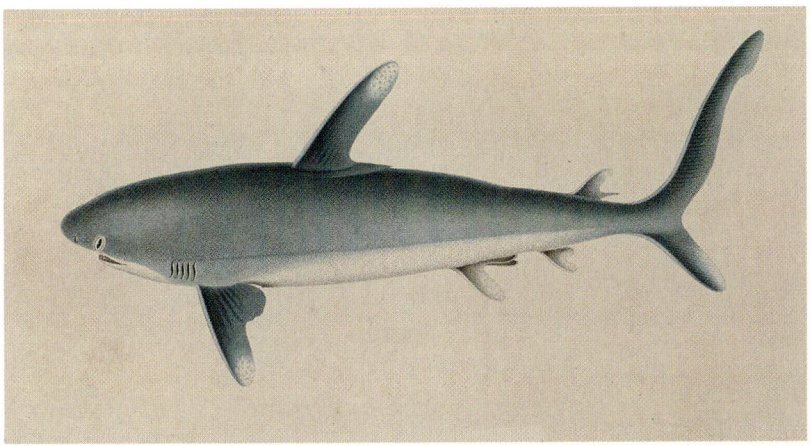

Opposite: *Blue shark, illustration from Jonathan Couch,* A History of the Fishes of the British Islands, *vol. 1 (1868 edn).* Above: *Oceanic whitetip shark (*Squale maou*), illustration from Louis-Isidore Duperrey,* Voyage autour du monde . . . Atlas *(1826–30).*

breathe. They have eely, narrow bodies and look like they've been trodden on by someone wearing a wellington boot; a ridged oval shape on their back is a modified dorsal fin, which acts as a sucker and lets them fix themselves onto other animals. A drawing of one appears in the notebook of Jan Brandes, from 1788, when the Dutch clergyman and watercolourist was living in Java. He painted the remora next to an oceanic whitetip shark and an albacore tuna, both species that can acquire an entourage of remoras. Mark Catesby painted and etched a remora, calling it 'The Sucking Fish'.

Mariners also told stories of an unusual type of octopus that they claim to have seen during their ocean-crossing journeys. Argonauts appear in Jules Verne's 1870 novel *Twenty Thousand Leagues Under the Sea.* The renegade mariner, Captain Nemo, and the captive marine biologist, Professor Aronnax, rise up in the submersible, the *Nautilus,* and watch all around them a shoal of hundreds of argonauts floating on the surface. Verne's book was a work of fiction but it was in part a popular-science book and included much of the latest marine-biological information

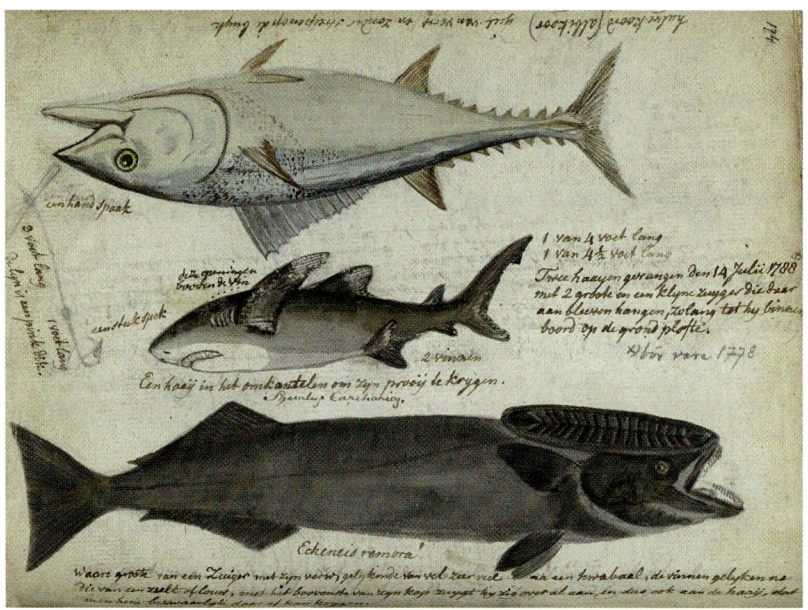

Jan Brandes, Sketch of Remora, Oceanic Whitetip Shark and Albacore Tuna, *1788, pencil and watercolour.*

available at the time. The book explains that argonauts (belonging to the genus *Argonauta*) are cephalopod molluscs, closely related to squid and cuttlefish, and that they have shells that they can come all the way out of – all true. Verne's story strays further from what we know today about these little animals when he describes how they use their shells as little boats and propel themselves across the waves by hoisting two of their arms into the air and spreading them wide like sails to catch the breeze. For an hour, the *Nautilus* cruises in the midst of these floating molluscs, then all of a sudden the argonauts are gripped by fear and decide it is time to dive: 'As if at a signal, every sail was abruptly lowered; arms folded, bodies contracted, shells turned over by changing their center of gravity, and the whole flotilla disappeared under the waves.'

Illustrated editions of the book often show Nemo and Aronnax gazing at the floating argonauts with their flattened arms in the air like hundreds of small ears. But we mustn't accuse Verne of making up this story. Wind-powered argonaut locomotion was a theory that many scientists at the time supported, fuelled nevertheless by mariner's tales.

American poet Marianne Moore (1887–1972) came closer to the truth about argonauts and how they use their shells in her poem 'The Paper Nautilus' (argonauts also go by this alternative name because their shells look superficially like those of their distant cousins, the chambered nautilus). While many of her other poems focus solely on the natural world, here Moore compares the fragile shell of the argonaut to writers who, despite gaining little recognition, can still find their own protective world in which to create works of great integrity and meaning. The 'dull/ white outside' of the argonaut shell shields a smooth interior that is as 'glossy as the sea'. The delicate shell holds something precious and profoundly beautiful. The argonaut uses her energies to attend to and cultivate the life, and art, being nurtured privately within. Its guardian, Moore says, 'scarcely eats until the eggs are hatched'.

The argonaut shell is not a boat, but a protective shelter. It had been French scientist Jeanne Villepreux-Power who worked this out. In the 1830s she studied argonauts from the Mediterranean, keeping them in her purpose-built aquarium tanks, and determined that they use the silvery webs at the end of two arms not as sails but to make their thin, papery shells. Only the female argonauts can do this. The males are shell-free and tiny in comparison to the females, which would fit neatly in the palm of your hand. For a long time, scientists mistook male argonauts, and in particular their detachable sperm-laden arms, as parasitic worms. Females make shells and use them as mobile brood chambers, to nurture their unhatched eggs. While many other octopus species live close to the seabed and lay their eggs in small caves and crevices, argonauts spend their lives swimming through the open seas and carry their eggs with them. Female argonauts will also sometimes swim to the surface and trap a bubble of air inside their shells. Then they swim a short way down, propelled by jets of water squirted from their siphon tube. The air bubble helps to keep

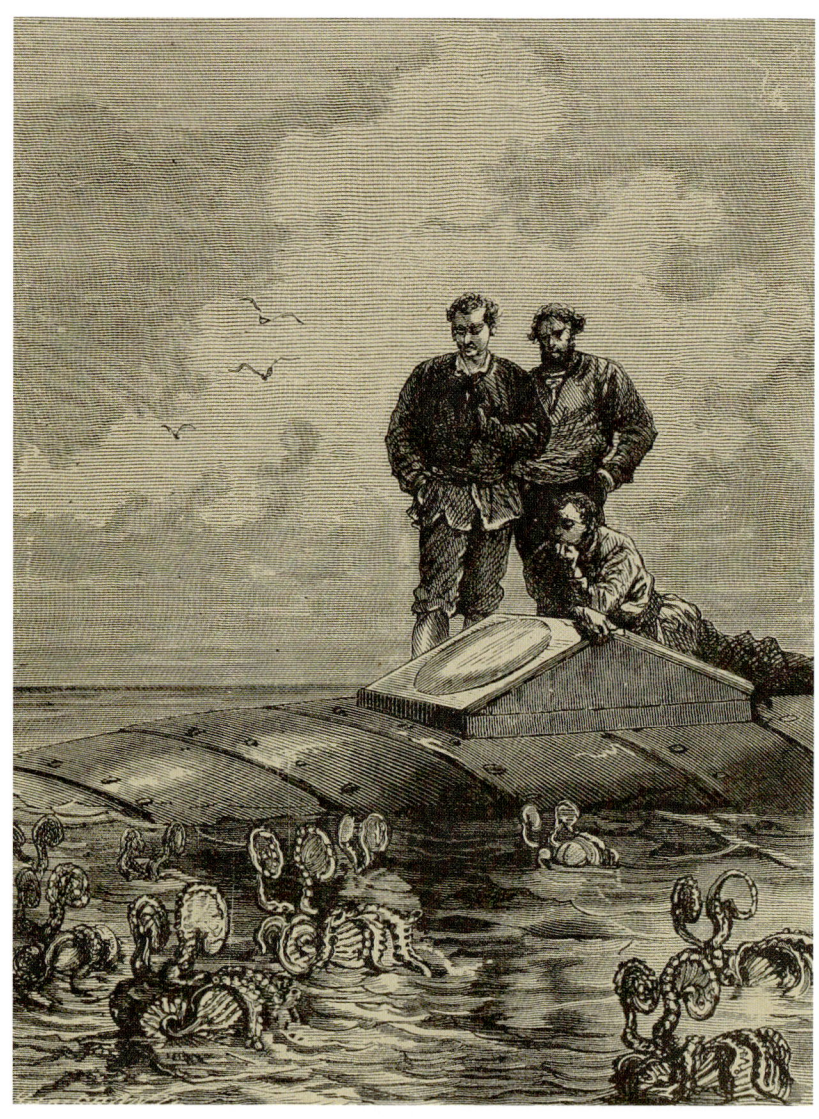

Above: *'A shoal of argonauts'*, *engraving from Jules Verne*, Twenty Thousand Leagues Under the Sea *(1875 edn)*. Opposite: *The Argonaut, or paper nautilus (*Argonauta argo*), from Giuseppe Jatta*, I Cefalopodi viventi nel Golfo di Napoli *(1896)*.

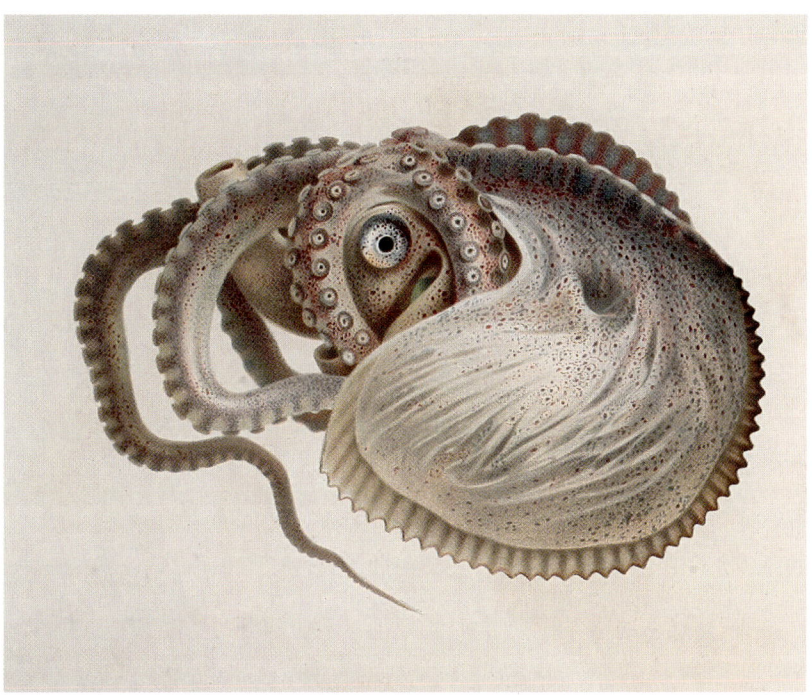

them buoyant and stops them from sinking too deep, allowing them to swim with ease just below the waves.

When a female argonaut dies, her delicate shell might float to the surface and only then is it likely to catch the breeze. It's a rare treasure to find an intact argonaut shell that has blown all the way onto a beach. Even more unlikely is to see a living argonaut close to shore. In 2012 a fisherman catching squid not far off the coast of Los Angeles caught a living female argonaut and brought her back to a local aquarium. The argonaut had likely been swept towards shore on a strong oceanic current. After a week in captivity, the female was joined in her aquarium tank by more than 20,000 babies, each one just a few millimetres in size, that had hatched from eggs nestled inside her shell.

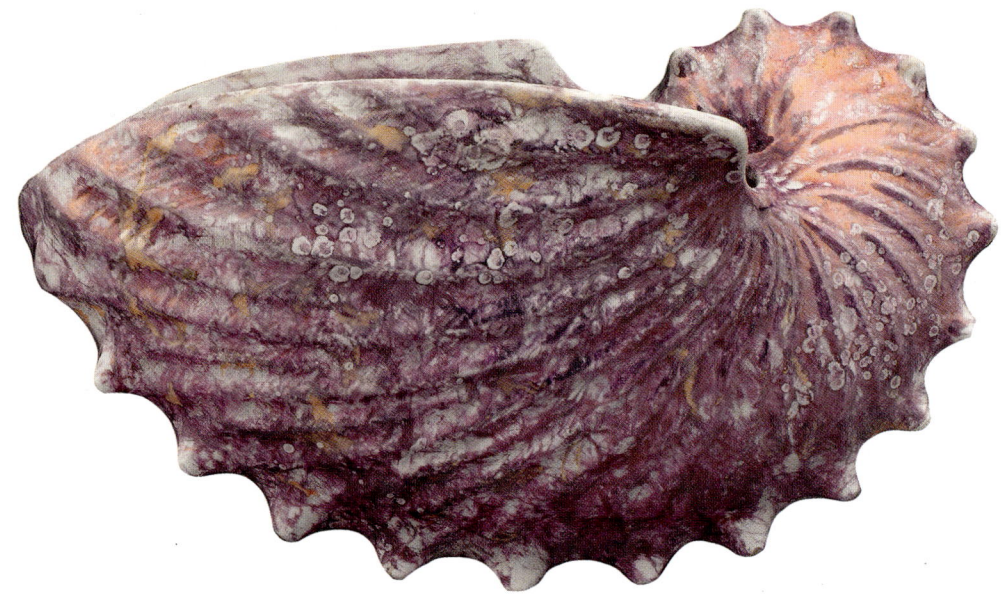

Argonaut wall pocket, Wedgwood manufactory, c. 1805, earthenware.

Glass Animals

Argonauts appear among the work of a father-and-son team who had the ambition of capturing the ephemeral beauty of the ocean's most delicate and intricate creatures. Leopold Blaschka was born in 1822 into a family of glassmakers in Bohemia. 'The only way to become a glass modeler of skill', he later wrote, 'is to get a good great-grandfather who loved glass.' Indeed, Leopold did inherit his ancestors' passion for glass and they taught him the skills of lampworking, now called flameworking, which involves using a flame to melt glass rods and tubes, then shaping them by hand. Leopold joined the family business making test tubes, costume jewellery and glass eyes, but those are not what he would be remembered for.

He married and the couple had a son, but Leopold tragically lost both his new baby and his wife in a cholera pandemic in 1850. A few years later, following the death of his father, the grieving

Leopold went on a recuperative ocean voyage to the United States. In the mid-Atlantic, near the islands of the Azores, the wind died and the sailing ship was becalmed and made little progress. In that time, Leopold contemplated the open sea around him and gazed at and sketched the intricate marine life that came drifting by. For some time after that, the translucent creatures of the open sea would hover in his subconscious and eventually helped inspire him to try something different in his glasswork.

Back in Europe, Leopold married again and had another son, Rudolf, and in his spare time he began making glass models of flowers. These caught the attention of the Duke of Montbazon, a royal patron of the arts and keen amateur botanist, who commissioned Leopold to make him a hundred glass orchids and other tropical plants for the displays in his palace in Prague. An exhibition of these model flowers was seen, in turn, by Ludwig Reichenbach, director of the Natural History Museum in Dresden, who had an idea of something else Leopold could make.

At the museum, Reichenbach faced a great problem of how to display boneless sea creatures to the public. Birds, mammals and reptiles are comparatively easy to transform into lifelike taxidermy displays. It is possible to stuff fish, but the only option for preserving the ocean's invertebrates – the jellyfish, sea anemones, octopuses and so on – is to soak them in formalin and alcohol. And as anyone familiar with museum collections knows, pickled marine invertebrates are, generally speaking, not at all beautiful. Their colours are sapped away by their confined ocean of preserving chemicals. Soft-bodied animals lose their form and shape, and tend to collapse into amorphous, gooey blobs.

Reichenbach wondered whether Leopold's glassworking could be the solution he was looking for, and commissioned a dozen models of sea anemones. They were to accompany an exhibition at the museum of colour lithographs of sea anemones by Philip Gosse. Leopold agreed and set about creating his first marine invertebrates made of glass. He used glass, resins, paint, wires and adhesives to carefully copy a selection of sea anemone species from Gosse's paintings. They were a huge hit with museum visitors, who were primed to adore these underwater wonders, this being the mid-nineteenth century when the ocean and aquariums were still the height of fashion and stirred tremendous public interest across Europe.

Delighted with the public's response, Reichenbach encouraged Leopold to quit making glass eyes and jewellery, and turn his attention full-time to making glass sea creatures. This idea blossomed into a business that flourished for decades, all through the remainder of Leopold's life and that of his son, Rudolf, who soon became his glassworking apprentice.

The timing for their enterprise was perfect. Museums were opening up in many countries and, like Reichenbach in Dresden, their directors were desperate to have their own beautiful displays of ocean species. Universities used them as teaching materials for their students, the glass specimens being much more useful to study, hold in their hands and look at from all directions, compared to the squishy preserved specimens.

These were not just scientific and practical items, but artworks of great beauty that showed the stunning, living wonders

*Glass model of a sea anemone (*Bunodes balli*), by Leopold and Rudolf Blaschka, late 19th century.*

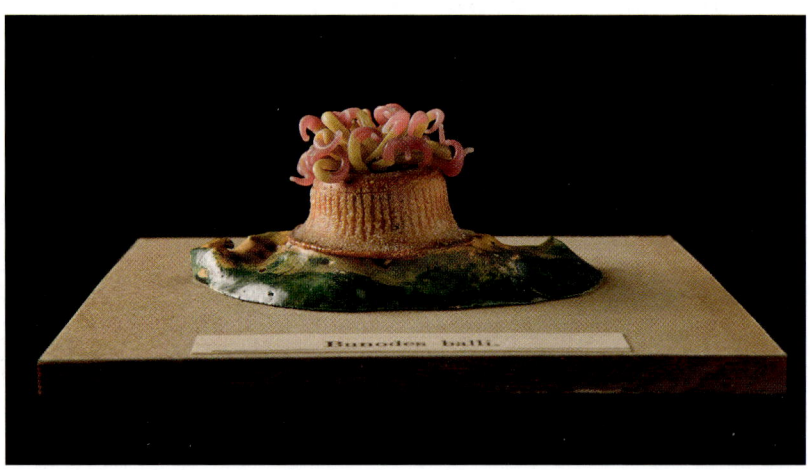

of the sea in three dimensions. Demand for Blaschka models came from private collectors, including those who preferred to replace real animals in their aquarium displays with glass animals that didn't require great skill to keep alive and looking attractive – they just need to be occasionally dusted.

Leopold and Rudolf set up a hugely successful mail-order business. After some smashed disasters, they worked out how to carefully pack their glass specimens to survive international shipping. Having begun with flower-like sea anemones the Blaschkas went on to recreate all sorts of other species that live throughout the ocean from the coasts to the open seas. Among the most spectacular are many kinds of diaphanous animals with fragile, translucent bodies. The Blaschkas made many different species of jellyfish that look as astoundingly intricate as the living animals themselves. Their flouncy, ruffled tentacles look like they're wafting in a gentle current; their smooth, rounded umbrella bodies look like they've been frozen in a moment between contractive pulses that push them through the water. As Ludwig Reichenbach had anticipated, glass was the ideal material to recreate the gossamer bodies of marine invertebrates like these. The texture and translucence make them come alive. And like the real creatures, you can see inside the bodies of glass jellyfish. Leopold and Rudolf gave their models all the right muscles and reproductive organs.

Early on the Blaschkas used illustrated books as reference material for their glass model making, then realized they needed to observe and sketch living species in three dimensions. So they did what many other artists and collectors were doing and set up aquarium tanks in their workshop.

Rudolf also returned to his sketches and memories of the sailing trip across the Atlantic. He would have encountered a unique ecosystem that exists at the surface of the open seas. Known as the pleuston, this floating assembly of animals inhabits the interface between seawater and air, and drifts across the ocean pushed by currents and winds. Many of these species evolved to be blue to match their surroundings and hide from predators and prey.

Leopold and Rudolf created many key species of this floating ecosystem, including many creatures that are easily mistaken for true jellyfish (scyphozoans) but are actually different types of closely related animals that all belong to the same phylum, the Cnidaria, as sea anemones and corals, and share characters like stinging tentacles. The Blaschkas made models of by-the-wind sailors (*Velella velella*) and blue buttons (*Porpita porpita*), which both have circular blue bodies and short blue tentacles. The by-the-wind sailors have the additional feature of a little triangular sail sticking up, which actually does what people thought argonauts do and catches the breeze to propel them across the surface of the open ocean. These two jellyfish-like animals belong to the cnidarian class Hydrozoa, and their bodies are made up of colonies of hundreds of individual units, called zooids. Different types of zooid perform specific tasks; some zooids defend the colony with stings, some catch food and digest, and some are reproductive and make eggs and sperm.

Other members of the sea surface pleuston ecosystem that the Blaschkas made are the Portuguese man-of-wars, which float at the sea surface with an inflated balloon of gas, dragging their long, blue and purple tentacles through the water beneath them to catch small fish. In these models, in particular, Leopold and Rudolf made splendid use of glass to brilliantly recreate the different types of zooid that make up the colony. There are dark blue coiling tentacles that catch food (called tentacular palpons), shorter pale-coloured fingerlike extensions for digestion (gastrozooids) and yellow clusters of reproductive units (gonozooids).

The Blaschkas completed the blue-tinted, floating open-ocean ecosystem with a type of sea slug known as the blue dragon or sea swallow (*Glaucus atlanticus*), which feeds on the stinging tentacles of Portuguese man-of-wars, by-the-wind sailors and blue buttons. Nobody has yet worked out how they do it, but somehow the blue dragon sea slugs don't trigger the stinging cells

*Glass model of a siphonophore (*Physophora hydrostatica*), by Leopold and Rudolf Blaschka, late 19th century.*

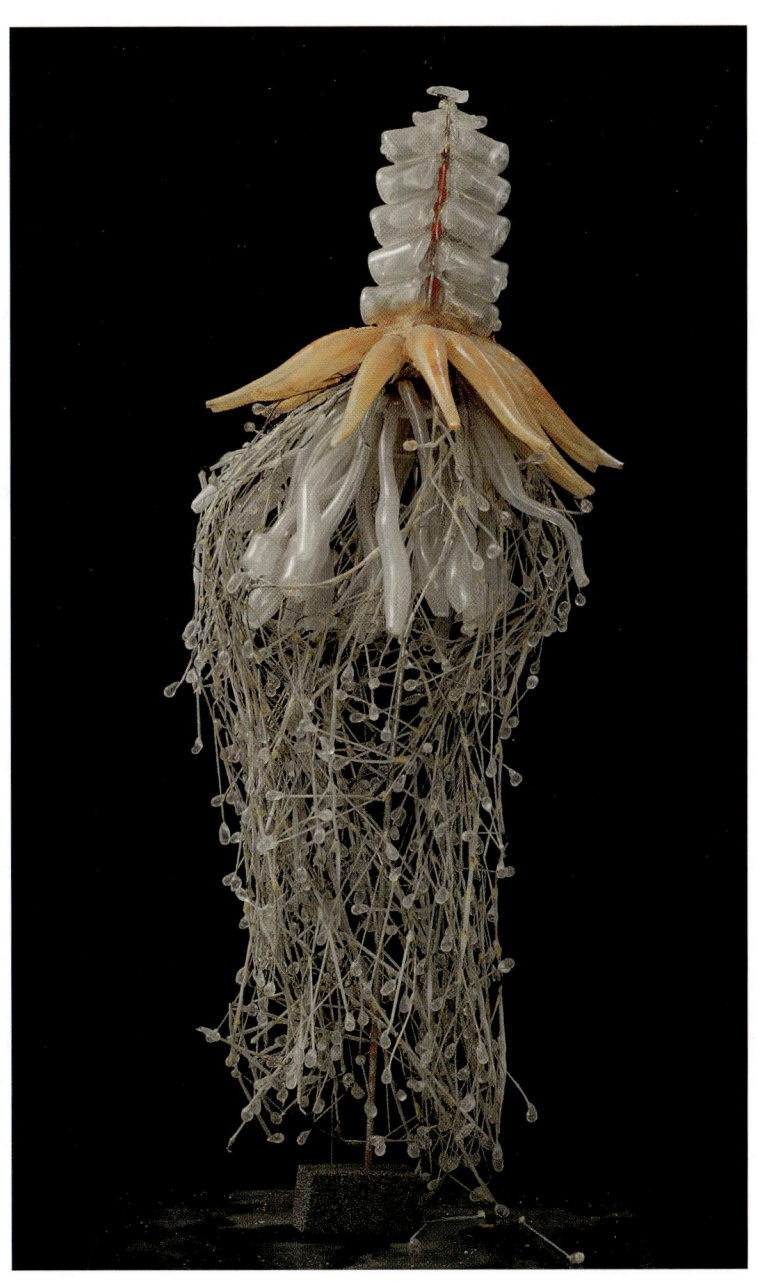

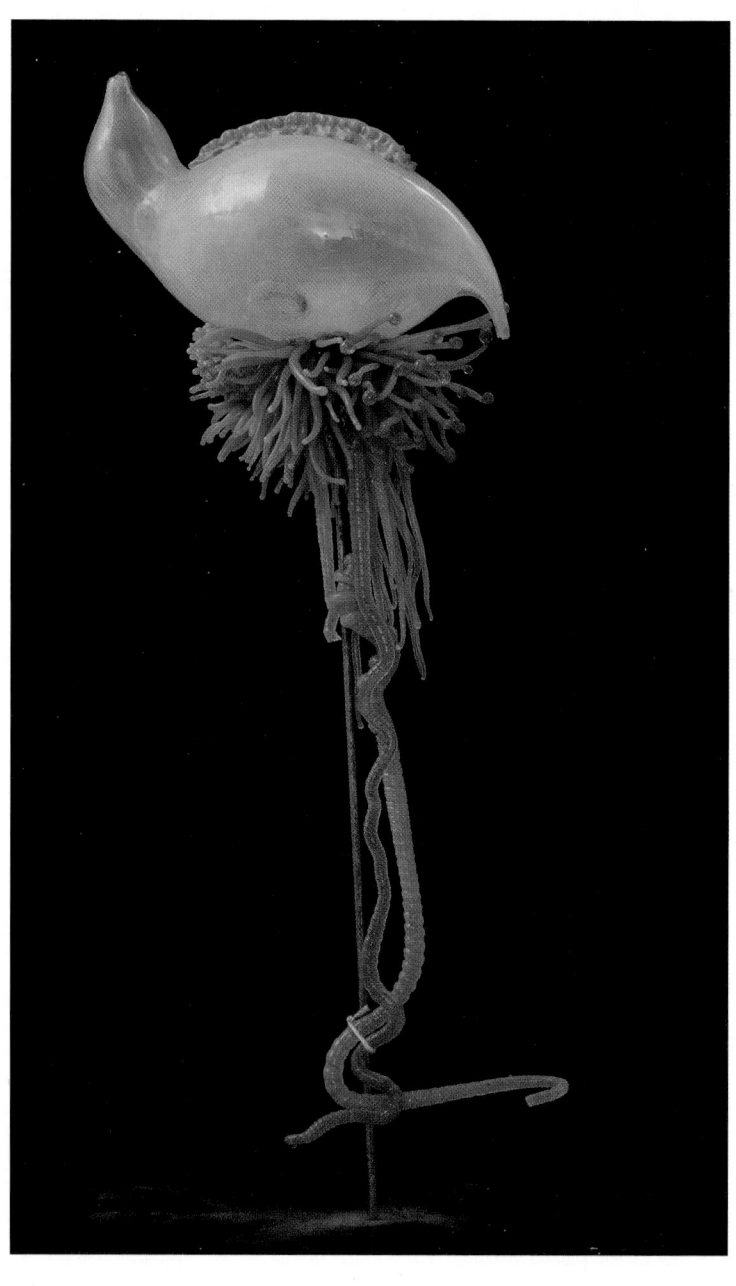

when they eat them, but keep them intact and push them into their long, finger-like projections and use the stings for their own defence.

Throughout their lifetimes, Leopold and Rudolf were relentlessly prolific. In their workshop in Dresden, Germany, the two men produced glass models of more than seven hundred species of marine animal – octopuses, squid, worms, starfish, sea cucumbers – and more than 10,000 individual sea creatures, which then swam off to populate a global glass ocean held in museums, universities and private collections around the world.

Rudolf chose not to train an apprentice and so the generations of glass-loving Blaschkas came to an end when he died in 1939. As technologies have moved on and photography and other forms of image making and scanning were developed, the technical uses for their glass models fell by the wayside and many of the collections were stored away and forgotten about. More recently, there has been renewed interest in the Blaschkas' glass sea creatures, as well as thousands of glass flowers they later made for the Harvard University Botanical Museum. Museum curators in many countries have come across dusty old boxes and found inside the gleaming, delicate bodies of Leopold and Rudolf's marine invertebrates. Many are in need of restoration and conservation as the resins, paints and glues have been deteriorating over the decades. Contemporary glassworkers have said that they wouldn't be able to recreate the Blaschkas' glass models. While they understand the techniques that Leopold and Rudolf used, nobody now has the nimble-fingered skills necessary to reproduce these intricate animals.

Hundreds of the Blaschkas sea creatures are once again on display to the public in museums worldwide, including the Natural History Museum in London, the National Museum of Ireland, Canterbury Museum in Christchurch, Aotearoa (New Zealand) and the Field Museum of Natural History in Chicago.

*Glass model of a Portuguese man-of-war (*Physalia pelagica, *now known as* Physalia physalis*), by Leopold and Rudolf Blaschka, late 19th century.*

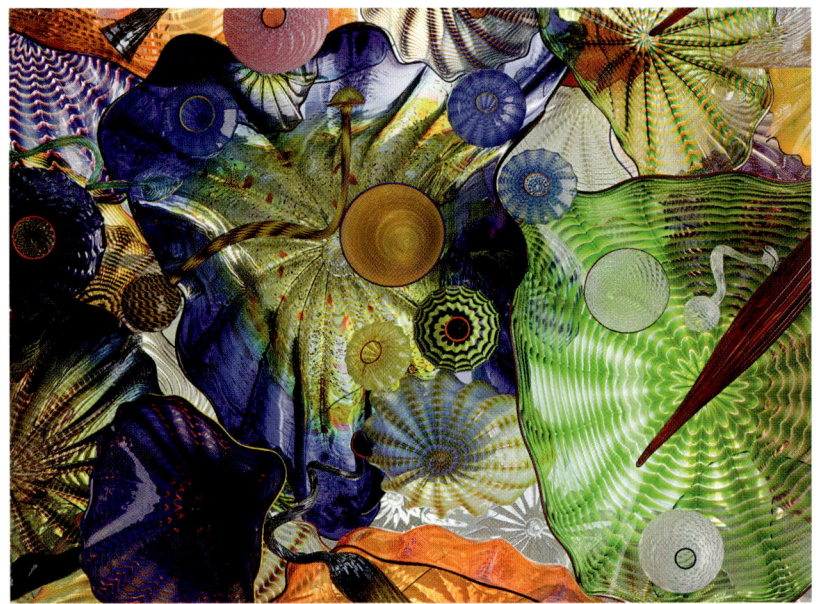

Glass ceiling sculptures by Dale Chihuly inspired by sea creatures, 2002, from the Seaform Pavilion of the Chihuly Bridge of Glass, Museum of Glass, Tacoma, Washington.

Occasionally a model comes up for sale to private collectors, such as a Portuguese man-of-war that sold at auction in London in 2019 for £10,000.

Other artists have used glass to capture the essence of pellucid sea creatures, notably American artist Dale Chihuly, known for his large glass installations of flowers and coiling tendrils. His series from the 1980s titled *Seaforms* encompassed thousands of pieces inspired by sea creatures. Although not recognizable as particular species, they vividly reflect on underwater life, its shapes, textures and movements, the way light passes through a sea anemone's tentacles or jellyfish's body.

Miniature Ocean Worlds

The most beautiful work by Leopold and Rudolf Blaschka that I have seen was a few years ago at the Natural History Museum in London. Through the magnificent entrance of this cathedral to the natural world, past the skeleton of a blue whale suspended overhead, is the Cadogan Gallery, which displays a small selection of the museum's greatest treasures. There's the stuffed body of an extinct great auk, a first edition of Charles Darwin's *On the Origin of Species*, the seven-hundred-year-old skull of a royal lion kept at the Tower of London and a piece of Moon rock.

The museum's curators also bring out the most stunning examples from their collection of Blaschka models. When I was there, on display was a sphere covered in spikes. As I remember it was roughly the size of a bowling ball. Most of the Blaschkas' models were life-size – but not this one. The creature this model is based on is almost invisible to the naked eye. This is a radiolarian, a type of single-celled, amoeba-like organism that floats through the open seas as part of the microscopic zooplankton. Aptly, radiolarians make an exoskeleton (or test) out of silica, the key ingredient of glass.

Even if I had been allowed to pick up the Blaschkas' radiolarian and turn it over in my hands, I wouldn't have dared for fear of breaking off one of the needle-thin extensions. Nevertheless, standing in front of its glass display case, walking around, looking from every angle, I came to appreciate how unique Leopold and Rudolf's works were. These are pieces of art and science combined, and they were meant to be viewed in person and seen in three dimensions. They were intended to help people imagine what sea creatures are truly like, whether that's an octopus sitting with their arms splayed across the seabed, or a minute living cell wafting through the sea.

The Blaschkas came to create models of minute plankton thanks to the work of their contemporary, the German zoologist, philosopher and artist Ernst Haeckel. Born in 1834, Haeckel trained to be a doctor in Berlin, to please his parents, but had little interest in medicine. He was much more enchanted by the natural world and art. In 1854 he spent a summer on the island of Heligoland in the North Sea with Johannes Müller, one of Germany's foremost scientists at the time and an enthusiastic marine biologist, whose work included classifying sharks. Müller inspired Haeckel to share his passion for ocean life and instructed him in the technique of towing a fine net through the water to catch minute plankton. After just a few weeks on the island, Haeckel wrote to his parents to tell them, 'my decision is made: in future, I shall become a naturalist, a zoologist in fact.'

Back in Berlin, Haeckel began to imagine a life for himself as a travelling naturalist and artist, exploring the world and chronicling the riches of nature with his paintbrush. Then he fell in love and realized that he must have a more stable career if he was to marry and provide for his sweetheart, Anna Sethe. His best option would be to seek employment at a university. For that to happen, he needed to produce his first published work, and so, at the age of 25 and with funds from his family, Haeckel set off for Italy to search for a research topic. He stopped in Florence to visit pioneering microscopist Giovanni Amici, and bought a powerful microscope that would let him magnify tiny creatures by up to a thousand times. Haeckel then spent a month in Rome enjoying its history and art, before heading to Naples. His plan had been to study the anatomy of starfish and sea urchins, but when he failed to find enough animals of the right kind he almost turned his back on science to become instead a landscape painter. Nevertheless, he persevered and spent six months studying sea life in the Strait of Messina, the narrow gap between the toe of Italy's boot and the island of Sicily – as the waters of the Mediterranean are squeezed between these two landmasses, they bring in a current rich in plankton, which became the focus of Haeckel's studies and fervour, in particular the radiolarians.

He gazed at drops of seawater and discovered a world of minute, hidden beings, and he began to draw. Here was a link that Haeckel had been struggling to find between the visual poetry of nature and a more scientific view of the world. His drawings and

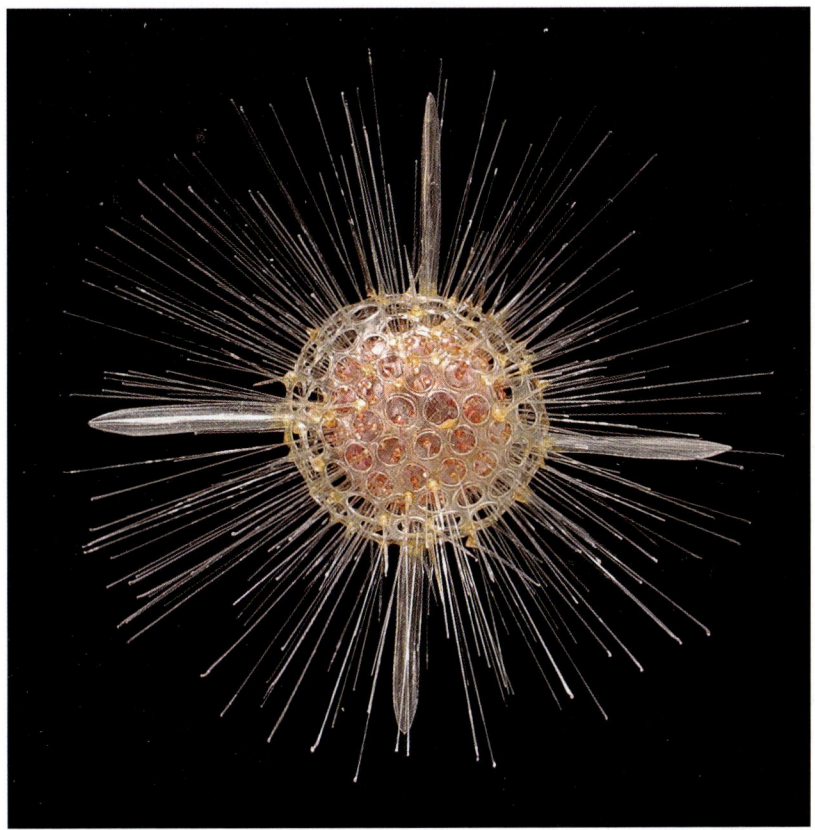

*Glass model of a microscopic radiolarian (*Actinomma asteracanthion*),
by Leopold and Rudolf Blaschka, late 19th century.*

paintings not only helped him understand the origins and lives
of living things but let him peer deeper at nature's beauty. He had
discovered a way to unite his loves of art and nature that would
also, he hoped, provide for Anna.

From Sicily, Haeckel wrote to his beloved fiancée to tell her
about his discoveries and how, on a single day of fishing with his
fine plankton net, he found twelve new radiolarian species, among
them the most 'charming creatures', as he put it. He wrote,

I knelt down in front of my microscope and cried my heartfelt thanks to the blue sea and the generous sea-maidens, the fair Nereids, who always bestow on me such glorious gifts. I promised to be good and decent, worthy of such fortunes and to dedicate my whole life to the service of glorious nature, truth and freedom.

In the Sicilian seas, Haeckel found what he needed to launch his career as a zoologist. He had discovered more than a hundred new species of radiolaria and after two more years' work he published *Die Radiolarien*, a two-volume monograph, which helped to secure him a professorship at the University of Jena.

The monograph features 35 plates, many in colour, of copperplate engravings based on hundreds of Haeckel's illustrations of radiolarians. He neatly arranged the radiolarians into symmetrical tableaus, a few different varieties on each page. Many are geodesic spheres that look wildly futuristic and strange, like designs for space ships and not like anything with a biological origin. Zoomed in so close to this microscopic view, it is easy to quickly forget that these are tiny sea creatures afloat in the huge expanse of the open seas.

In most of his images, Haeckel excluded the living parts of the radiolarians, partly a practical matter because the delicate tissues don't preserve well and so he wouldn't have seen them except in the very freshest specimens right after he netted them from the sea. While still alive, an indistinct miasma of living tissue can squeeze out through the perforated holes in the skeleton to grab passing morsels of food. The aspect of these creatures that captivated him most lay in the symmetrical and orderly structures of their skeletons – shapes that lead us back to the ocean. The long needles sticking from the glass orbs drastically increase their surface area and act as buoyancy aids, keeping the radiolarians afloat in the upper layers of the sea; there, some species soak up the sunlight with minute algae lodged inside their bodies, something biologists only learned of after Haeckel's time, although we can see these green blobs in some of his drawings.

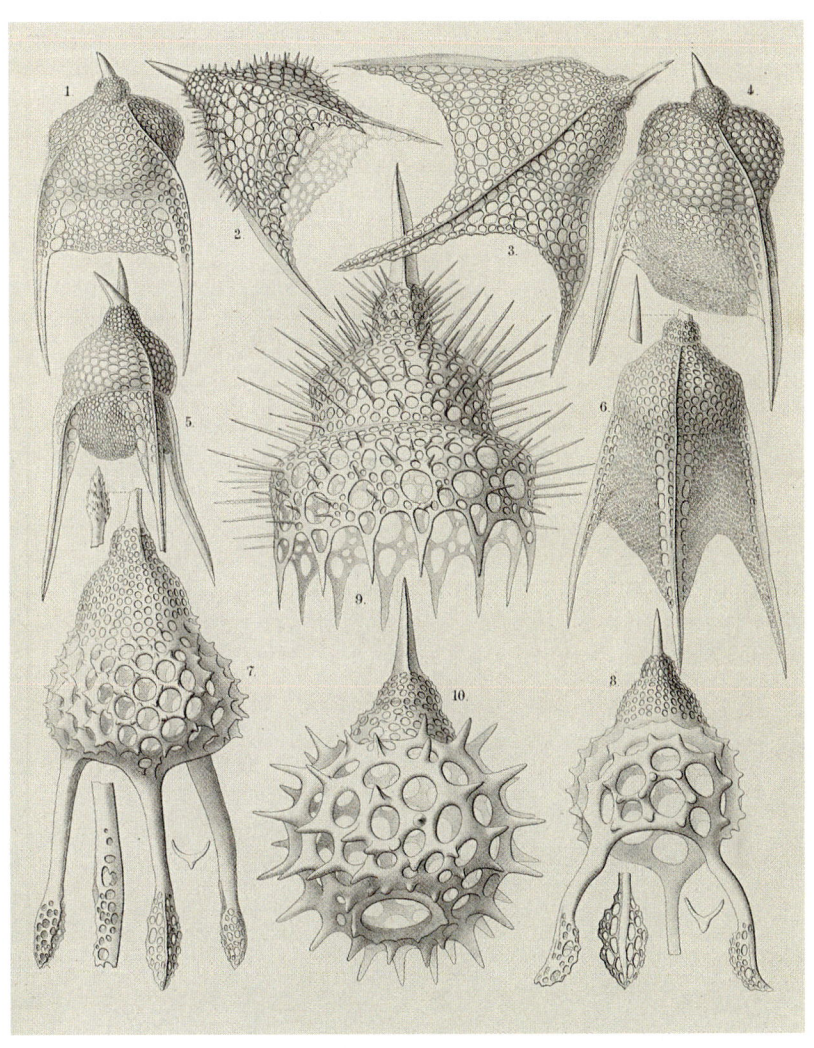

Above: *Drawing of microscopic radiolarians (*Rhizopoda radiaria*),
from Ernst Haeckel,* Die Radiolarien, *vol. 11 (1887).* Overleaf left:
Copepods, from Ernst Haeckel, Kunstformen der Natur *(1904).*
Overleaf right: *Marine worms, from Haeckel,* Kunstformen
der Natur.

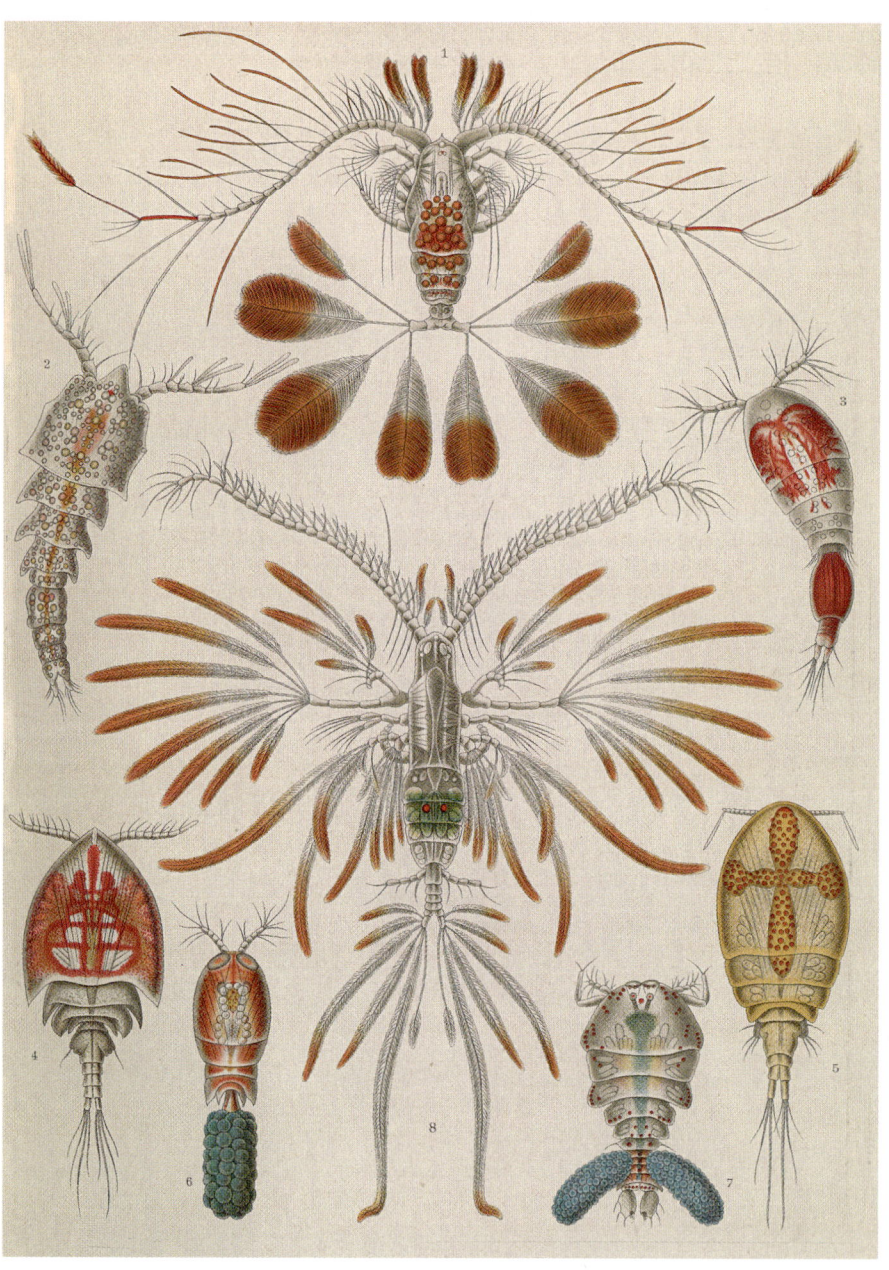

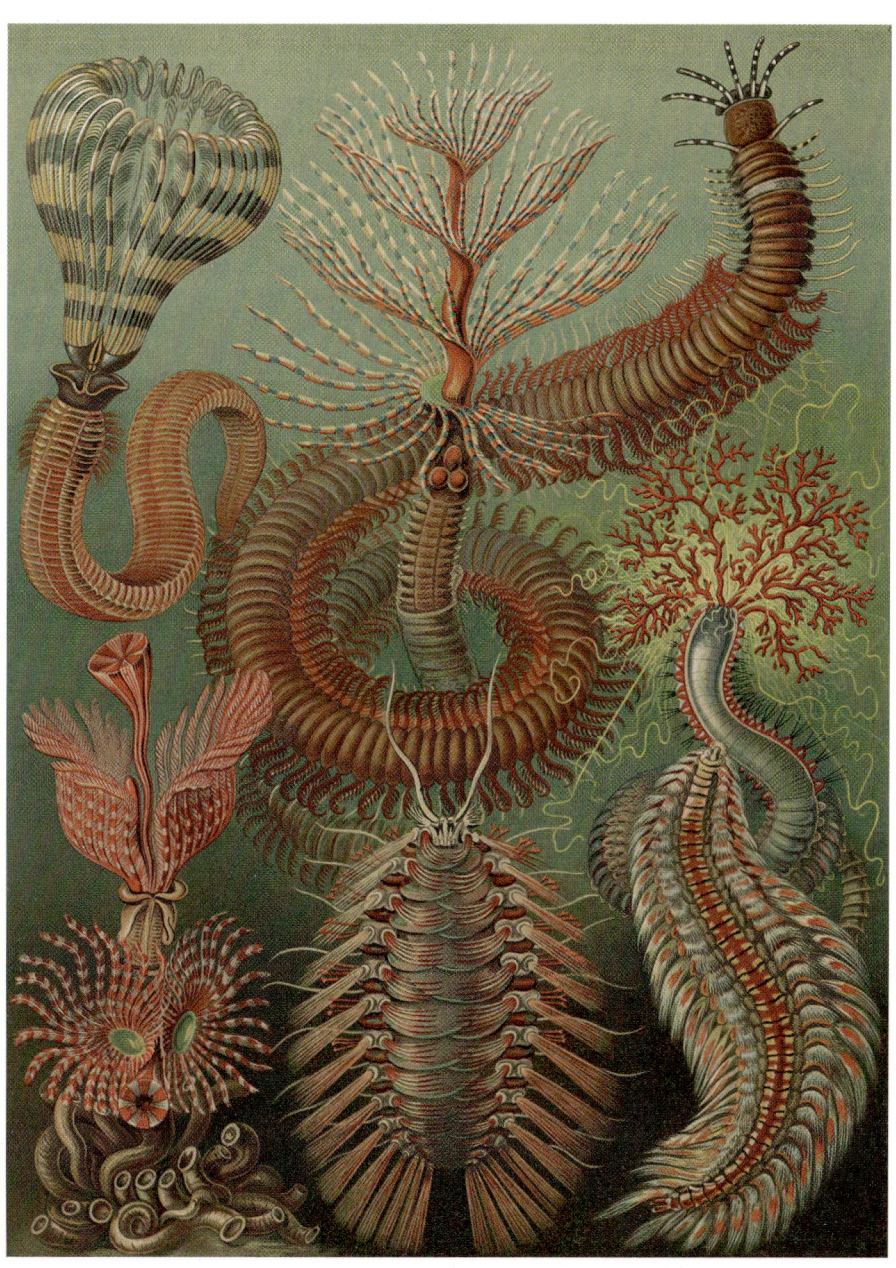

Haeckel's radiolarian monograph and drawings were the basis of his academic career as a scientist. To him, drawing and painting were tools for gathering knowledge about nature. His work showed that there is incredible beauty to find in the smallest facets of nature. Still today people gaze over his shoulder at the aesthetic wonders of this miniature world. Haeckel has inspired scientists to look at nature in different ways, and he inspired many other artists too.

In Germany, he had been friends with Leopold and Rudolf Blaschka. They corresponded and exchanged ideas, and he lent them illustrated books from his library to help them create their glass models, especially the plankton. It's a deeply subjective matter whether the Blaschkas' models were works of science or art, or both. Today the models are mostly owned and displayed in natural-history and science museums, although they would equally take their place in art galleries.

What is clear, however, is that Haeckel left a tremendous impression on art, architecture and design. He saw this for himself in 1900 when he visited the Exposition Universelle, the World Fair, in Paris and walked through a gigantic, ornate metal gateway. The French architect René Binet said that everything about his design for this *porte monumentale*, from the tiniest details to the overall shape, had been inspired by Haeckel's studies and drawings of radiolarians. The 1900 world fair is considered the high point of Art Nouveau, a movement that pushed against the machine-driven, excessive industrialization of the times and instead drew inspiration from nature. Haeckel found himself at the centre of this new way of looking at the world, largely thanks to another series of drawings he had begun publishing, and for which he remains remembered today.

Kunstformen der Natur (Artforms in Nature) is a series of one hundred colour prints that Haeckel published in sets of ten from 1899 to 1904. Each page is devoted to a group of organisms – most of them sea creatures – based on his sketches and watercolours. These were a true blend of art and science, with the species artfully arranged on the page, specifically designed to appeal to the general public. In these pages Haeckel took viewers to many habitats around the world, and he led them down into the most inaccessible region of the ocean. He showed that the deep waters of the abyss are not an empty void, as many had believed them to be, but in fact inhabited by unique and elegant creatures.

FOUR

THE ABYSS

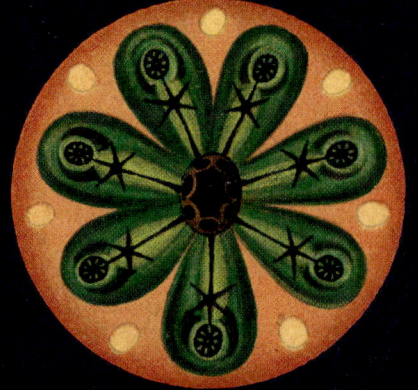

The ocean's greatest depths remain the most mysterious and little-known parts of the planet. This space is the furthest from human lives as it's possible to reach, geographically, physically and psychologically. The only equivalent to this inhospitable, faraway place is outer space, and there are many parallels to draw between these two realms. Since the very earliest times, humans have turned their minds up to the skies and down into the ocean deeps and wondered what lies in those places and what greater meaning might be found in both.

Unlike the star-filled heavens that anyone on a dark night can look upwards and see, the deep ocean is firmly out of sight. Until the development of deep-diving technologies over recent decades, the deep sea has been completely out of bounds for people. This is simply somewhere we cannot be.

Out of the human inaccessibility to the deep, two chief responses have emerged. One has been for people to designate the hidden deeps the ideal domain of monsters and powerful deities, places inscrutable, foreboding and so far out of reach that it is impossible to go and check what is really going on down there. A second reaction to the deep ocean has been to assume that there's nothing down there at all and that this vast, dark realm, with crushing pressures and freezing temperatures, is too extreme for any life to exist. As a result, cultures have interpreted the deep ocean with different ideas and images of what could lie far below. Many people have made this place a breeding ground for whimsical stories and fairy tales, and imaginations have run wild. Then there are the curious-minded explorers and scientists who invent ways of venturing into the depths, either in person or using remote tools, to learn about this faraway realm. And the more people have investigated life thousands of metres beneath the waves, the more the real beasts of the deep have shown themselves to be just as strange and wonderful – sometimes even more so – than any imagined monsters.

Adventures and Monsters

Scientists define the deep ocean as everything that lies below the 200-metre (656 ft) mark, and from there on down they've encountered and described thousands of unearthly looking life forms and habitats. Before we get to those explorations, including returning to the work of Ernst Haeckel, there are alternative views of the abyss to consider that came from people who have defined the deep in different terms. Rather than beginning at a specific depth, the deep sea can be framed as any place that is out of reach, where fishing lines and nets don't easily stretch and divers cannot plunge. The inaccessible depths and their mysteries are the ideal settings for stories that imagine what it would be like to sink into the deep and experience an aquatic life.

The outrageous adventures of the fictional German soldier Baron Munchausen have been told in illustrated books, films and comic strips. The original stories were written in the 1780s by the German writer Rudolf Erich Raspe, who is said to have based the baron on the real-life soldier Hieronymus Karl Friedrich von Münchhausen, who was renowned for telling tall tales of his exploits (in the 1950s his name was given to a mental health condition in which somebody pretends to be ill so that others will care for them and give them attention). In the Munchausen stories, the absurd, comedic character gets into all sorts of scrapes, travelling to the Moon, wrestling a 40-foot crocodile and exploring the underwater world on the back of a web-footed horse, surrounded by astonished, gawping fish, as depicted in various illustrations. The baron, sporting his three-cornered hat, also gallops around underwater in the 1962 Czech film *Baron Prášil*, directed by Karel Zeman and known in English as *The Fabulous Baron Munchausen*. The highly stylized film is part live action and part animation, inspired by the engravings of Gustave Doré from

Baron Munchausen riding his web-footed horse underwater, illustration by J. B. Clark, from Rudolf Erich Raspe, The Surprising Adventures of Baron Munchausen *(1895).*

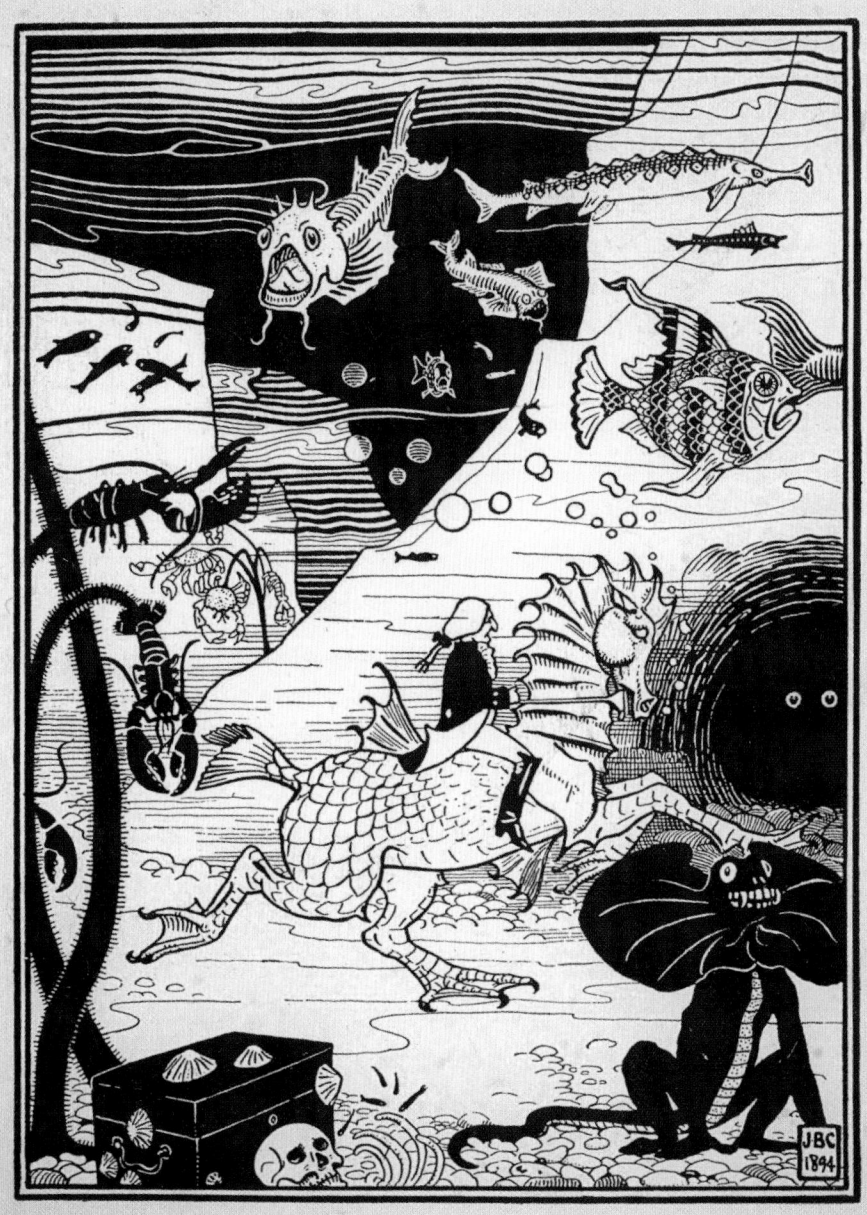

nineteenth-century Munchausen storybooks. Munchausen gets swallowed by an enormous fish only to appear again, whole and intact, at the other end of the fish's body, which ends in a second enormous mouth.

The early twentieth-century British artist Warwick Goble illustrated many books of fantasy, science fiction and folk tales, where his flowing style of watercolour painting lent itself perfectly to underwater scenes and diaphanous sea creatures. In a book of Japanese fairy tales, Goble painted a monkey dressed in fine robes, riding through a deep underwater seascape on the back of a giant jellyfish. He also illustrated a 1909 edition of Charles Kingsley's children's classic *The Water Babies*, the story of a chimney sweep called Tom who falls into a river, drowns, turns into a water baby and has adventures with other water babies and aquatic creatures. In 1916 the renowned American illustrator Jessie Willcox Smith

Below: *'The jellyfish takes a journey', illustration by Warwick Goble from Grace James,* Green Willow and Other Japanese Fairy Tales *(1910).*
Opposite: *Underwater scene illustration by Warwick Goble, from Charles Kingsley,* The Water-Babies *(1909).*

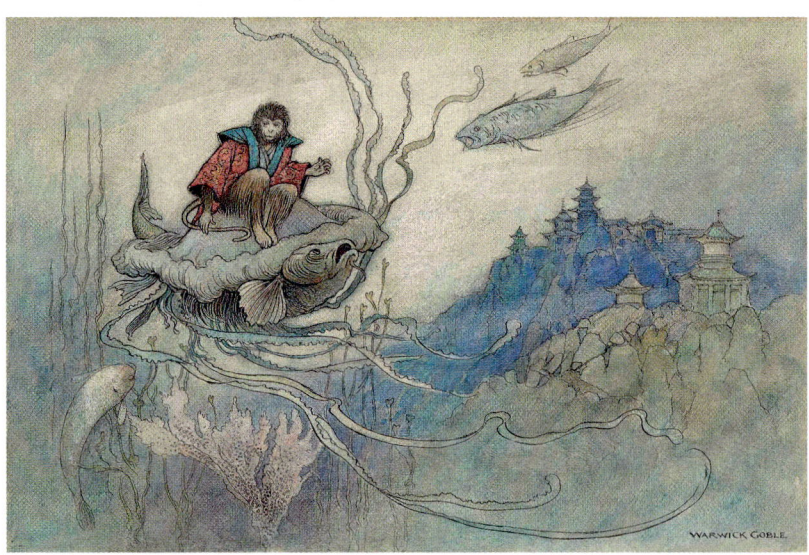

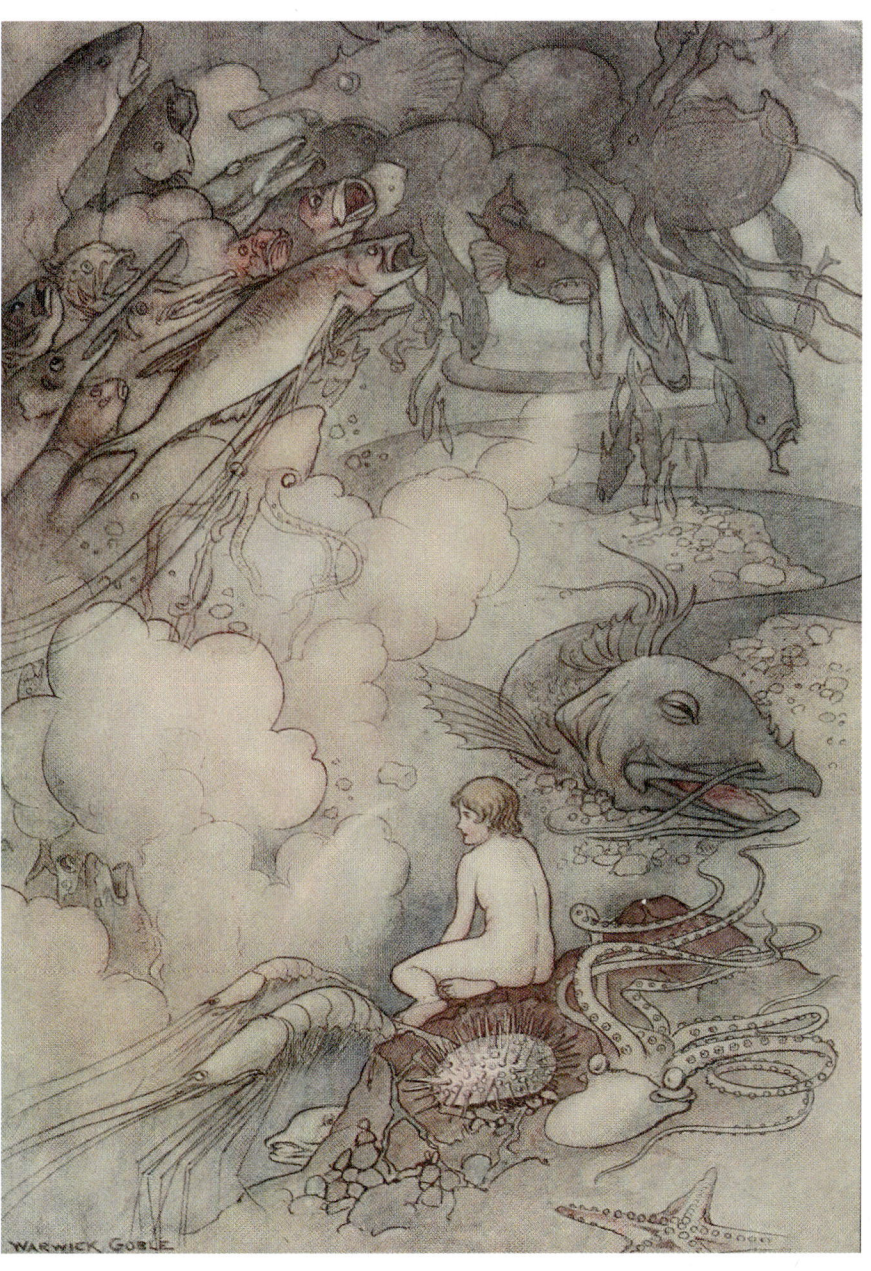

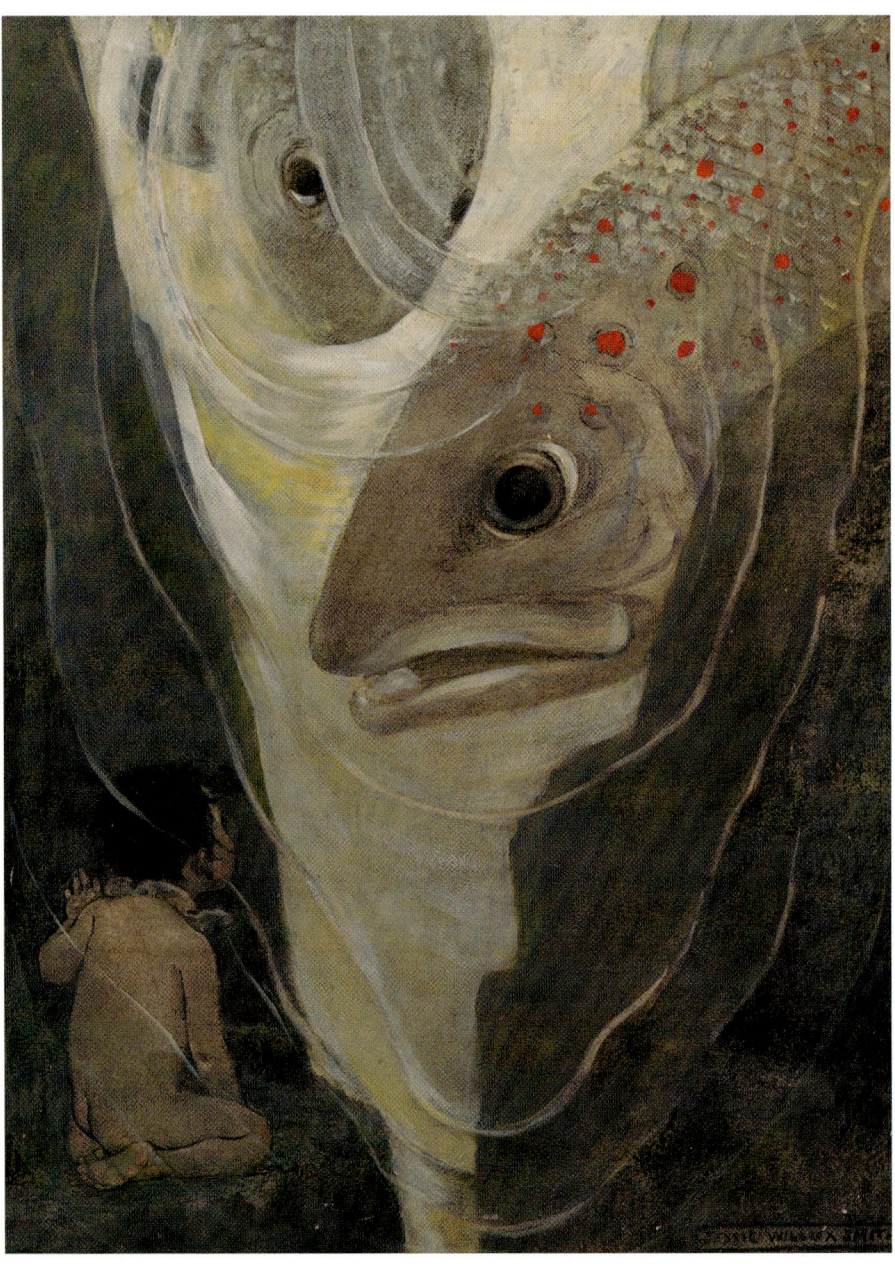

painted a scene from the book in which two giant fish peer at Tom the water baby.

A classic story that imagined what it would be like to live in the abyss is French author Jules Verne's *Twenty Thousand Leagues Under the Sea*. Verne placed his characters inside a futuristic submarine, the *Nautilus*, which took them anywhere they wanted to go in the ocean, no matter how far beneath the waves. He fitted out the crew of the *Nautilus* with deep-sea diving equipment that let them walk around on the seabed, to harvest food and materials to make clothing, and to mine underwater coalfields.

Illustrations from early editions of the book include the French artists Alphonse de Neuville and Édouard Riou's 'A Walk under the Waters', showing three divers striding through an aquatic scene that echoes Christian Schussele and James Sommerville's *Ocean Life*. De Neuville and Riou also imagined what the teardrop-shaped *Nautilus* would look like as it pierced through a shoal of thousands of squid and fish, and they gave the giant octopus a spine-chilling, wide-eyed glare directed towards the terrified men inside Captain Nemo's submerged vessel. The aesthetics of Verne's deep-sea explorations went on to inspire the steampunk subgenre of science fiction, with its retrofuturistic technologies. His stories of Nemo remained popular into the twentieth century, including the 1954 Walt Disney film adaptation *20,000 Leagues Under the Sea* directed by Richard Fleischer and featuring James Mason and Kirk Douglas, with the climactic scene of a fight with that most iconic deep-sea creature – a giant squid. Building the animatronic squid was one reason why the film went way over budget and was allegedly up to that time the most expensive Hollywood production ever.

Opposite: *Two fish peer at Tom the water baby, scene from Charles Kingsley,* The Water-Babies *(1916), watercolour illustration by Jessie Willcox Smith.* Overleaf left: *The* Nautilus *submarine surrounded by a shoal of squid, engraving from Jules Verne,* Twenty Thousand Leagues Under the Sea *(1875 edn).* Overleaf right: *'A walk under the waters', engraving from Verne,* Twenty Thousand Leagues Under the Sea *(1875 edn).*

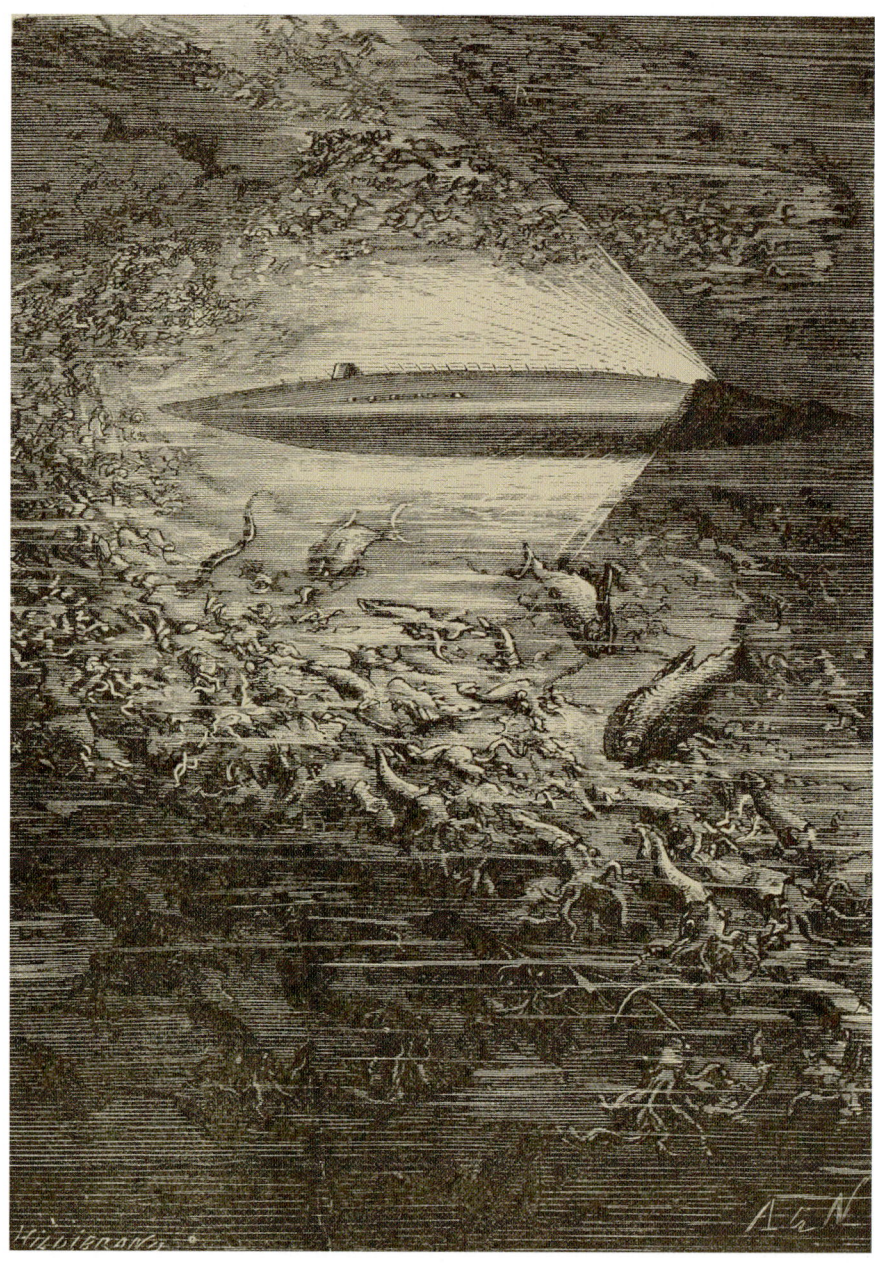

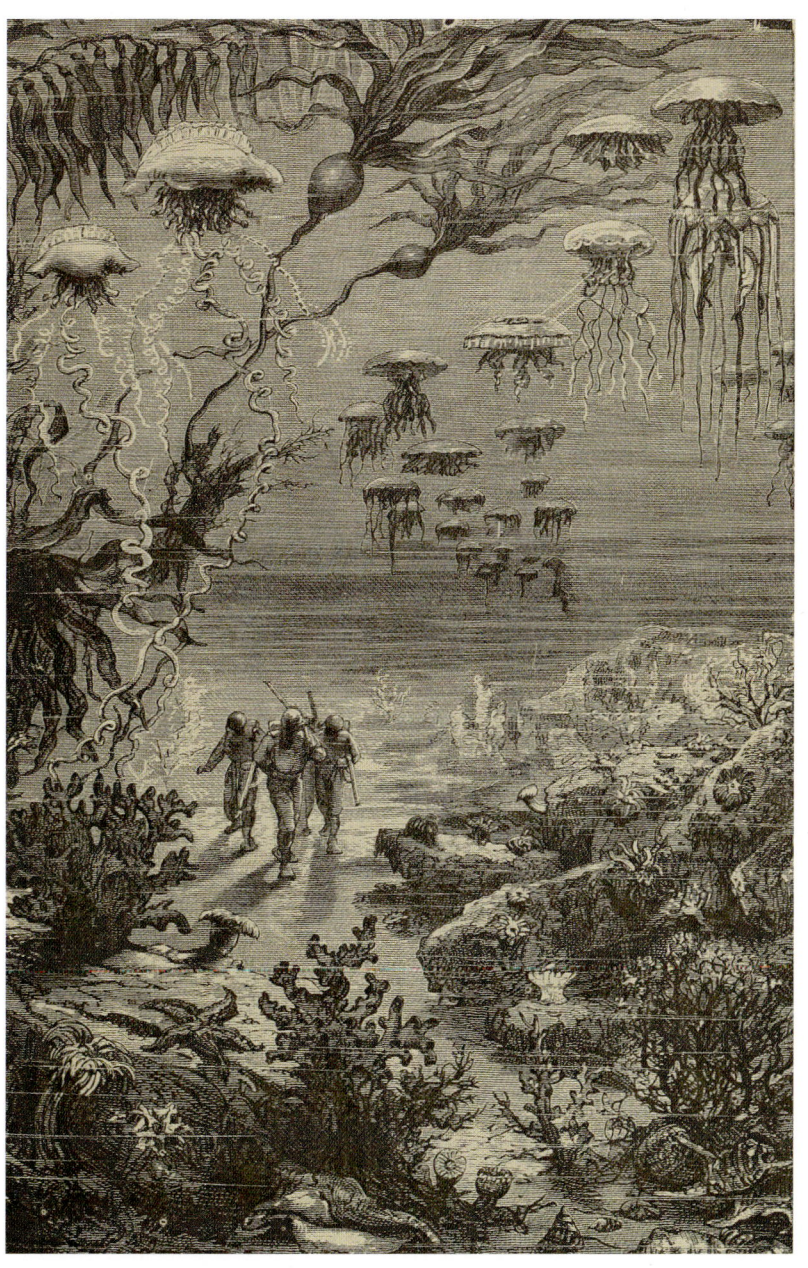

Countless historical drawings of giant squid show just how obsessed people have been about these enormous cephalopods and the reputation they've gained as the embodiment of terrifying monsters of the deep, such as the kraken. Some illustrations show the bodies of these animals washed up on beaches. An etching by an unnamed artist in 1661, *Zeemonster gevangen tussen Scheveningen en Katwijk* (Seamonster between Scheveningen and Katwijk), is an extraordinary image of a giant squid lying on a beach with its eight arms splayed out straight and stiff like a starfish, the snapping, dark bird-like beak in the centre of the head.

Illustrations commonly show giant squid attacking ships, wrapping their huge arms and tentacles about the deck, while sailors flee in panic. These scenes are, of course, utterly fictitious. Giant squid (*Architeuthis dux*) and their slightly shorter but heavier relatives, the colossal squid (*Mesonychoteuthis hamiltoni*), live in the twilight zone of the deep ocean, between 200 and 1,000 metres (655–3,280 ft) down. On rare occasions when they venture to the surface, they are sick and dying and in no fit state to pick a fight

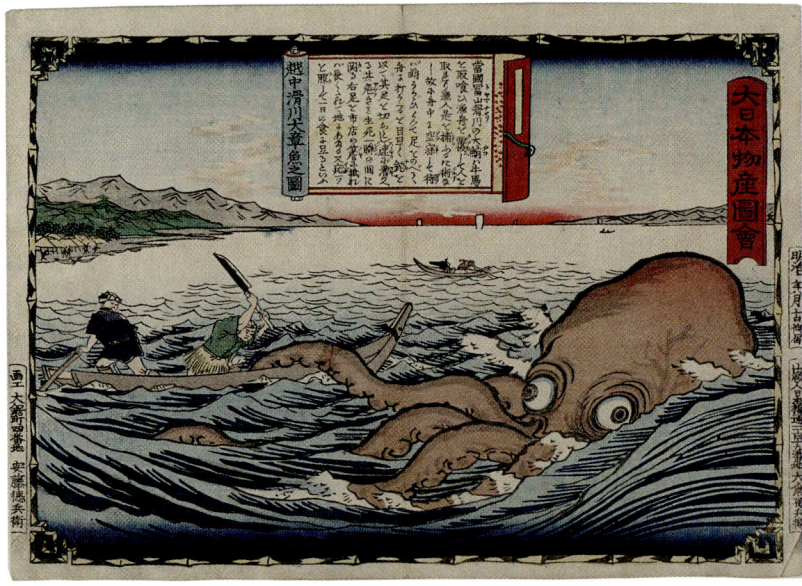

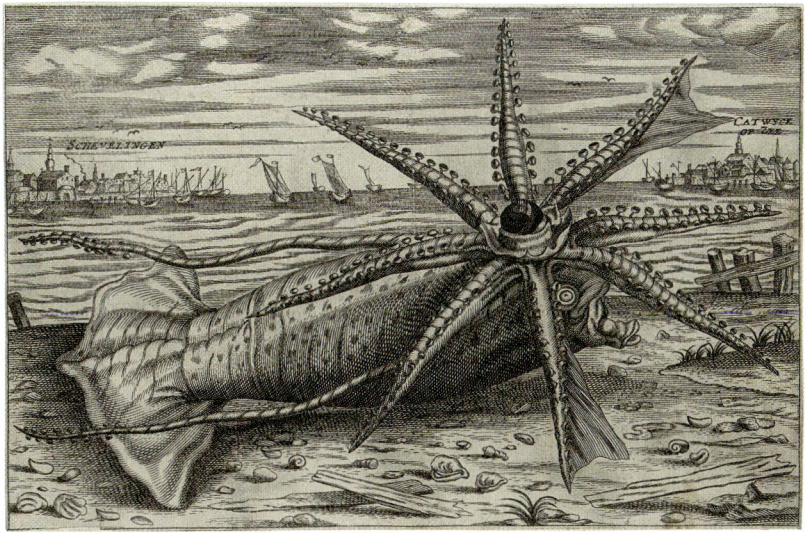

Opposite: *Utagawa Hiroshige* III, Hunting the Giant Octopus
of Namekawa in Etchu Province, *1877, colour woodblock print.*
Above: *Etching of a giant squid on a beach in the Netherlands, 1661.*

with people. They are big, for squid, this being a case of what is
known as deep-sea gigantism, which sees various species growing to abnormally large sizes in the ocean's depths. But they're not
so huge that they would be able to swamp a ship. At most, their
bodies are a few metres long.

Images that are more realistic, although no less dramatic,
imagine what a fight would look like between a giant squid and
a sperm whale. Remains of giant squid beaks in sperm whale stom-
achs and circular sucker scars on their skin show that whales and
squid certainly do have close-up encounters in the deep. No one
has actually ever seen this happening; live giant squid have only
been filmed in the deep sea a few times and never in the clutches
of a sperm whale. But studies using electronic devices fixed to
whales to track their movements show they do indeed chase after
squid at relatively high speeds, using creaking sounds to echolo-
cate and follow them through the darkness of the twilight zone.

Tentacles in the Deep

Cephalopods were important in early exploration of the deep ocean. They helped to thoroughly disprove the widely held assumption that the abyss is an empty, lifeless void. And they showed that not all animals in the deep ocean are grotesque, fanged monsters; there is much delicate beauty in the hidden depths.

The theory of a lifeless or azoic deep was popularized in the 1840s by a British naturalist, Edward Forbes, who found what he thought was solid evidence showing that life petered out with greater depths and was completely absent below 300 fathoms (roughly 550 metres/1,800 ft). Forbes studied the life of the Aegean Sea and used a dredging net to scoop up animals from the seabed. As his net came up, emptier and emptier, he was convinced that this was a pattern of diminishing life that would hold across the ocean. However, the Aegean isn't like other seas; it is sparsely populated due to a lack of nutrient inputs. Had Forbes looked elsewhere with suitable equipment, as many after him began to do, he would have seen that life in the ocean goes on and on.

Towards the end of the nineteenth century, the search for life in the abyss entered a new era when teams of scientists from Europe and North America embarked on long voyages of discovery. The sailing ships and steam-powered vessels were equipped with the latest research tools to probe the physical, chemical and biological secrets of the deep. They carried miles of piano wire to lower down with weights attached to measure the depth of the seabed, and powerful winches to pull dredging nets back up on deck. These research trips pre-dated deep-diving technologies that would allow humans to venture into the deep themselves. Animals were caught in nets and brought to the surface as carefully as possible, so that scientists on deck could observe them before their colours faded and shapes collapsed.

The reports from these expeditions opened a vivid window onto the living wonders of the deep as they had never been seen before. Many were accompanied by arresting illustrations, including reports from the Deutsche Tiefsee-Expedition (German Deep

Sea Expedition), now generally known as the *Valdivia* expedition after the ss *Valdivia*, the 90-metre-long (295 ft) former passenger ship that had been retrofitted for scientific research with laboratories and dredging gear. German marine biologist Carl Chun had proposed the trip, which was funded by the German government and set off from Hamburg in August 1898 with the aim of studying life deeper than 500 fathoms. Chun was the scientific leader of the expedition, which spent nine months navigating the Atlantic, Indian and Southern Oceans, a journey of 32,000 nautical miles. The *Valdivia* sailed as far north as the Faroe Islands, traced the west coast of Africa, spent a month dodging icebergs and pack ice near Antarctica, then swung northwards to Indonesia, India and the Seychelles, before travelling north along the Red Sea, through the recently opened Suez Canal into the Mediterranean, then back to Germany. The ambitious journey brought Chun and his team to many different climates of the surface seas, from the frozen pole to the tropics, and all the while their attention was focused on what lay far down beneath the *Valdivia*'s deck.

On the ship's return, Chun devoted himself to studying the specimens that they collected and writing up the *Valdivia*'s findings. The expedition had brought back so much material from the deep sea that it took numerous scientists forty years to compile 24 volumes of detailed reports. Chun died in 1914 at the age of 61; he didn't live to see the work finished, but he produced his own publications from the expedition that introduced to scientists and the general public the extraordinary living world that he and his team had uncovered. He was a teuthologist, an expert on cephalopods, and wrote the volume of the *Valdivia*'s reports dealing with the octopuses and, most importantly, a stunning array of deep-living squid.

As well as Chun's detailed written descriptions, the cephalopod volume, published in 1910, is illustrated with large-format colour lithographs by the zoologist and expedition artist Friedrich Wilhelm Winter. Until the development of deep-sea diving machines that can film and photograph animals in their element,

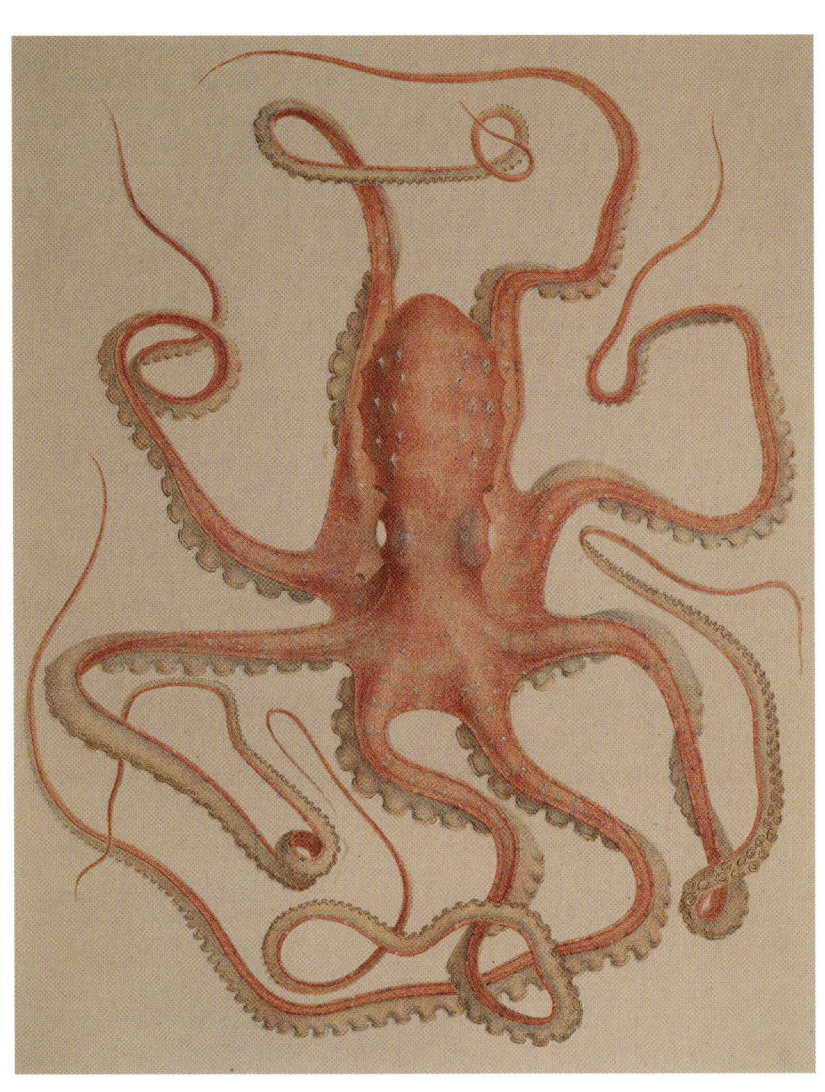

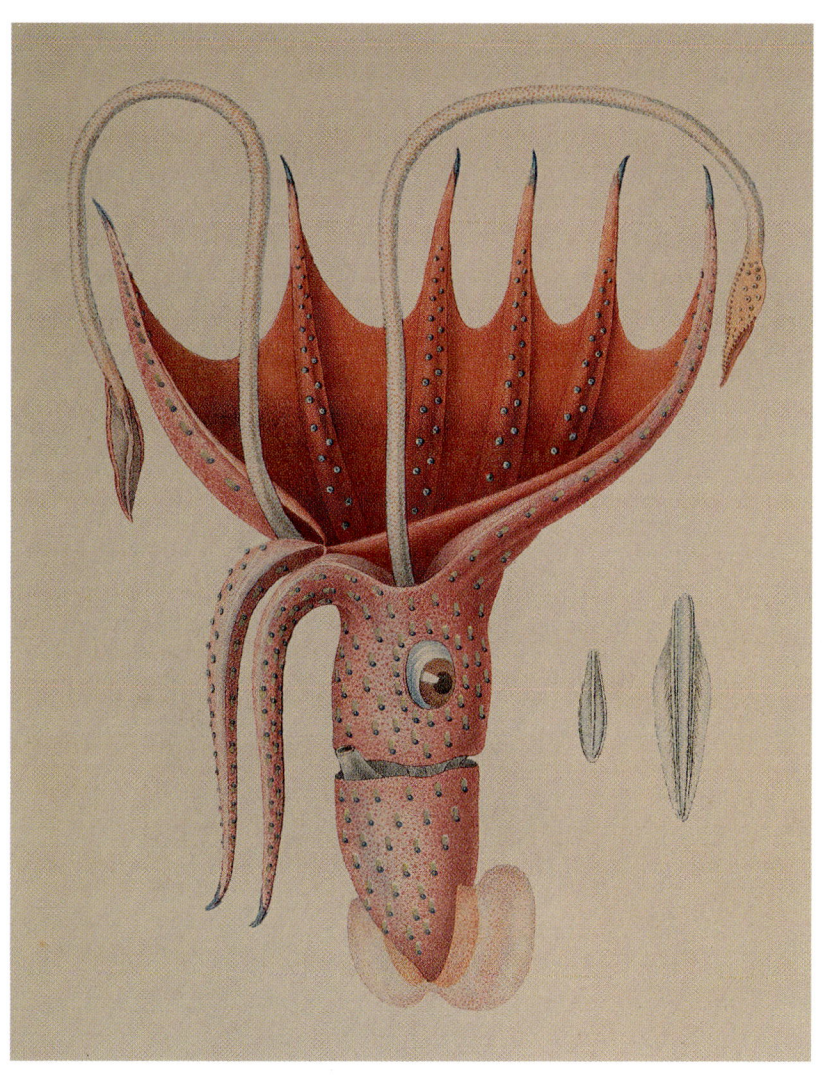

Opposite: *White spotted octopus, colour lithograph from Jean Baptiste Vérany*, Mollusques méditeranéens *(1851)*. Above: *Strawberry squid, colour lithograph from Baptiste Vérany*, Mollusques méditeranéens.

drawings like Winter's were the most accurate and beautiful portrayals of deep-sea animals. He would have seen the freshly caught animals on the deck of the *Valdivia*, sketched and observed them, then later created these colour lithographs. Preserved specimens brought up from the deep can never match the detail and aesthetics of these illustrations, especially for soft-bodied animals like the cephalopods.

In the UK, at the library of the Marine Biological Laboratory in Plymouth, with its ocean-view across Plymouth Sound, I take down from the shelves one of the surviving copies of the original *Valdivia* report and carefully look through many of the pages that were folded to give enough room for the huge drawings of life-size squid.

The plate of the jewel squid (*Histioteuthis bonnellii*) hasn't been coloured so I have to imagine for myself the purplish-red skin that this species has to help them hide in the shadowy depths, between 500 and 2,000 metres (1,640 and 6,560 ft) down. With no red wavelengths of sunlight penetrating this far, red pigments appear dull and indistinct, the ideal tactic for camouflage in the open waters of the deep where there's nothing to hide behind. The jewel squid illustration in the *Valdivia* report is minutely detailed. I can see the dots, known as photophores, that cover their body and give these squid their common name and light up like a twinkling gem, likely to distract attackers or perhaps to communicate with other squid. This one gazes out from the page, fixing me with a large eye. On the other side of its body, which I can't see, would have been a smaller eye, this being a member of the cock-eyed squid, a genus in which all the species have mismatching eyes. A study of the jewel squid's close relative, the strawberry squid (*Histioteuthis heteropsis*), found that the two eyes are for looking in two directions at once and deal with the contrasting light conditions of the twilight zone. One eye is regular-sized and blue, and it gazes down into the depths looking out for dim flashes of animals that glow in the dark. The other is yellow and twice as big. That one is good for looking upwards towards the brighter surface, watching for the silhouettes of predators and prey passing overhead.

A magnificent drawing of an emperor squid (*Chiroteuthis imperator*) graces another double, fold-out page of the *Valdivia* report. Chun, Winter and their team would have caught their specimen off the coast of Sumatra in Indonesia. The bodies of these squid grow to at least 30 centimetres (12 in.). Their arms and tentacles can reach more than 1.5 metres (5 ft) and in the drawing are neatly arranged around the squid.

The report contains smaller cephalopods, like the flapjack octopus (*Opisthoteuthis*), which would fit neatly in the palm of your hand. Video footage of these endearing little things, filmed in the wild by deep-diving research robots, has recently made people fall in love with them. They have flaps on the sides of their body, which look a bit like elephant ears but are not for listening; rather, they help them to swim short distances. They have short arms linked together by webs of skin that can billow out like a parachute as they float down towards the seabed, where they spend most of their time curled into a pancake-shaped lump, just as they are in the drawings in the *Valdivia* report. Winter's illustrations are amazingly accurate, given that they were produced from sketches of animals that had been caught and dragged hundreds or thousands of metres up from their native habitat.

Many of the animals in the *Valdivia's* cephalopod report are species that nobody had seen before. One discovery that Chun made was the angel octopus (*Velodona togata*), which he caught off the east coast of Africa from 749 metres (2,457 ft) down. These small, pale octopuses have large membranes between their arms, which can spread out and look like angelic wings.

Chun also found and named the vampire squid (*Vampyroteuthis infernalis*, meaning 'vampire squid from hell'), a name that conjures a terrifying apparition. In Winter's illustration the squid certainly looks ominous, with jet-black skin and blood-red eyes. Chun thought this was a close relative of the flapjack octopus, because they have similar webbing stretched between their arms and a pair of flaps on their body. In fact, these are not octopuses or true squid, but the only surviving member of the cephalopod family Vampyroteuthidae, and in reality they are quite gentle creatures.

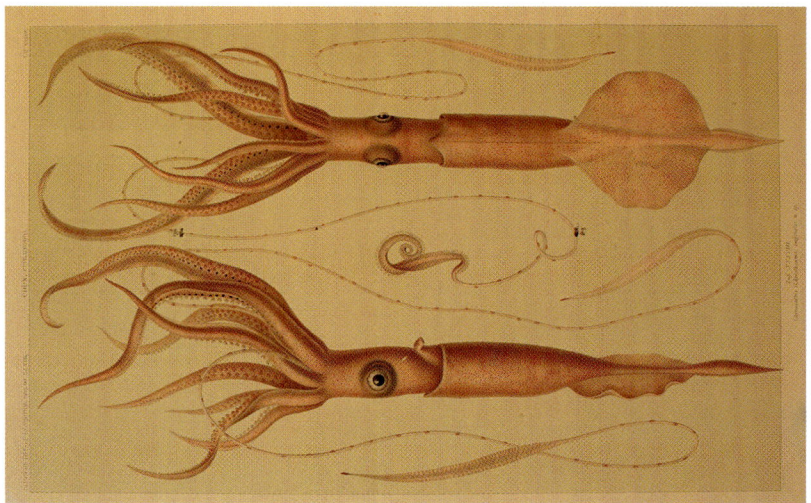

Above: *Emperor squid, illustration from the* Valdivia *expedition report by Carl Chun, vol.* XVIII *(1910).* Opposite: *Googly-eyed squid, illustration from the* Valdivia *expedition report by Carl Chun, vol.* XVIII *(1910).*

Octopuses and squid are active predators and chase after live prey, but the vampire squid are altogether more peaceful. Decades after Chun first found them, deep-sea scientists filmed them alive in the deep sea and confirmed that they feed themselves by catching flakes of marine snow that drift down from the sea surface.

While most of the illustrations in the *Valdivia*'s cephalopod report are of individual animals alone on the page, alongside enlarged details of parts of their body, a few are drawn swimming through the dark, deep sea. Chun found and named several new species of glass squid, which have rotund, transparent bodies and big eyes. One species, *Teuthowenia pellucida*, is aptly known as the googly-eyed squid. In the drawings, the glass squid's eyes shine brightly, due to the glowing photophores around them. This could either catch the attention of other glass squid, or alternatively hide their darkly pigmented retinas from attackers. Twilight zone animals face the common problem of predators looking up at them from below and seeing a dark silhouette against the brighter

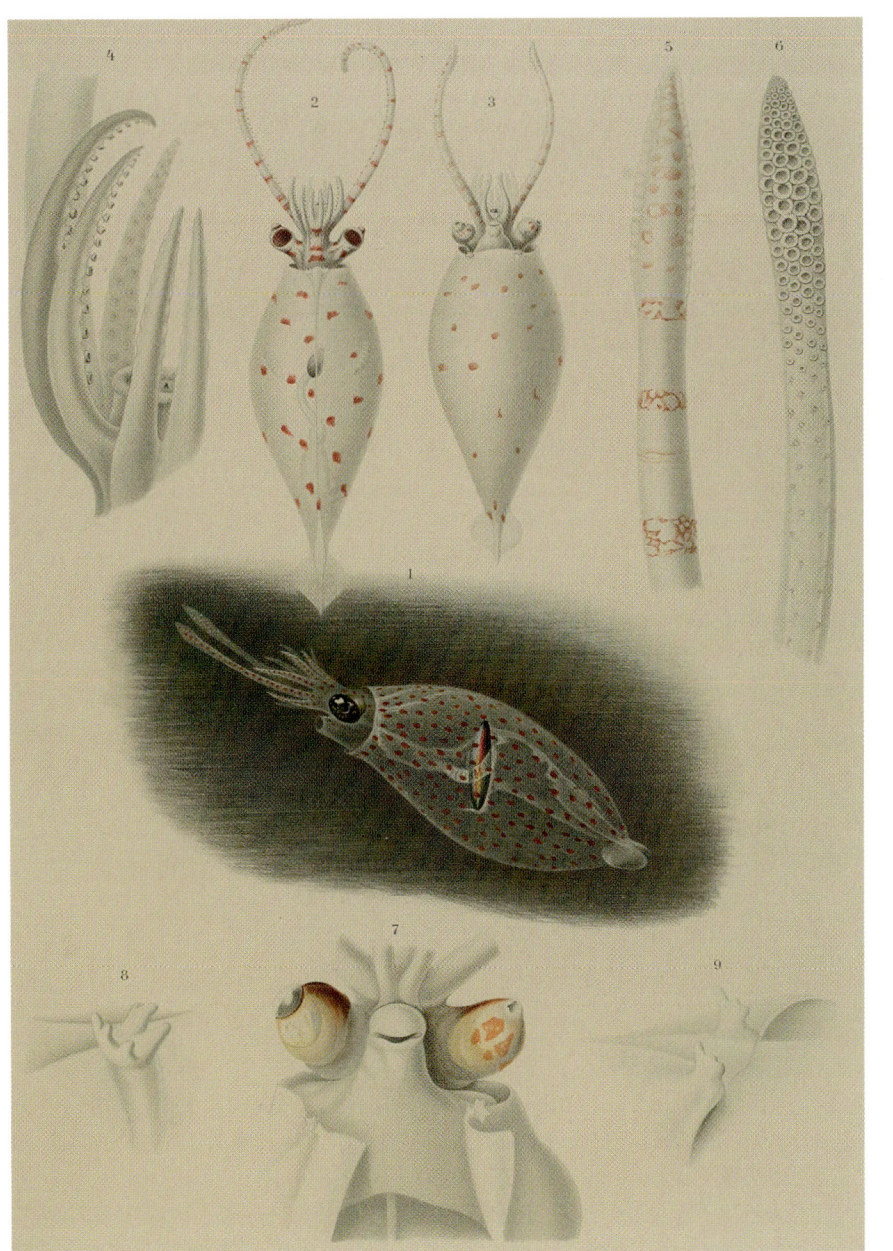

surface of the sea above. To get around this, many species illuminate parts of their bodies to precisely match the downwelling light and disguise their dark outline in the open waters of the twilight zone, where there's nothing to hide behind.

Lights and Spikes

The *Valdivia* expedition collected a lot more besides squid and octopuses. The fifteenth volume of the expedition's reports is dedicated to deep-sea fish, again illustrated by Friedrich Wilhelm Winter. The fish drawings reflect Carl Chun's aim of investigating the ways that animals adapt to living in this deep, dark realm. Among the illustrations, there are many of huge jaws filled with fearsome teeth, all the better for catching whatever prey comes along in the dark – in the deep sea, predators can't be picky about what they eat. Many of these fish have enormous eyes that help them catch the remnants of sunlight that trickle into the twilight zone and the brief flashes of light from bioluminescent animals. The fish themselves have lights dotted around their bodies. Some have glowing orbs on the end of forehead prongs or dangling from their chin, to act as a fishing lure to coax prey towards those huge mouths. The stoplight loosejaw (*Malacosteus niger*) has a pouch under each eye that uncommonly makes red light – most bioluminescent species make blue or green light, and occasionally yellow. Given that most deep-sea animals can't see red light, the loosejaws have evolved a secret wavelength that lets them see through the dark without being seen themselves, as if they are wearing night-vision goggles.

The colours of these illustrations show two contrasting ways in which fish hide in the deep sea. Ultra-black skin is common among deep-sea fish, including the dragonfish and anglerfish. Scientists recently found that deep-sea fish are some of the blackest animals on the planet, with skin as black as the darkest artificial materials constructed from carbon nanotubes. Being incredibly dark means the fish don't give themselves away by reflecting flashes of light from other bioluminescent animals or from their own glowing lures – there's no point tempting in prey if they can see the gaping mouth waiting for them. An alternative approach, as seen in hatchetfish, is to have a super shiny, silvery body combined with belly lights that can adjust their intensity and colour to match the light filtering down from above, disguising their silhouette.

Winter's drawings of fish show a tremendous range of body shapes that are unique to the deep among these animals that look quite otherworldly. The most alien-like, to me, is the young stage of the ribbon sawtail fish (*Idiacanthus fasciola*), a type of barbelled dragonfish. Winter's drawing shows the thin, wormlike body, with bulging eyes on the ends of great long stalks, like a pair of deely boppers, the popular head gear from the 1980s. A study in 1981 explored the vision of these early-stage fish, and showed that by holding their eyes so far apart, they can see almost a complete sphere of space around them, all the better for spotting prey and

Hatchett fish and lanternfish, illustrations from the Valdivia *expedition report by Carl Chun, vol. xv (1906).*

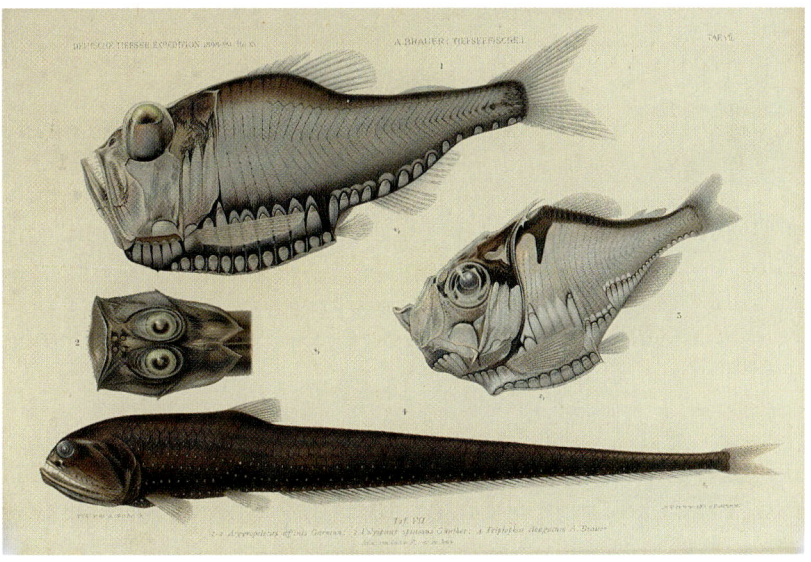

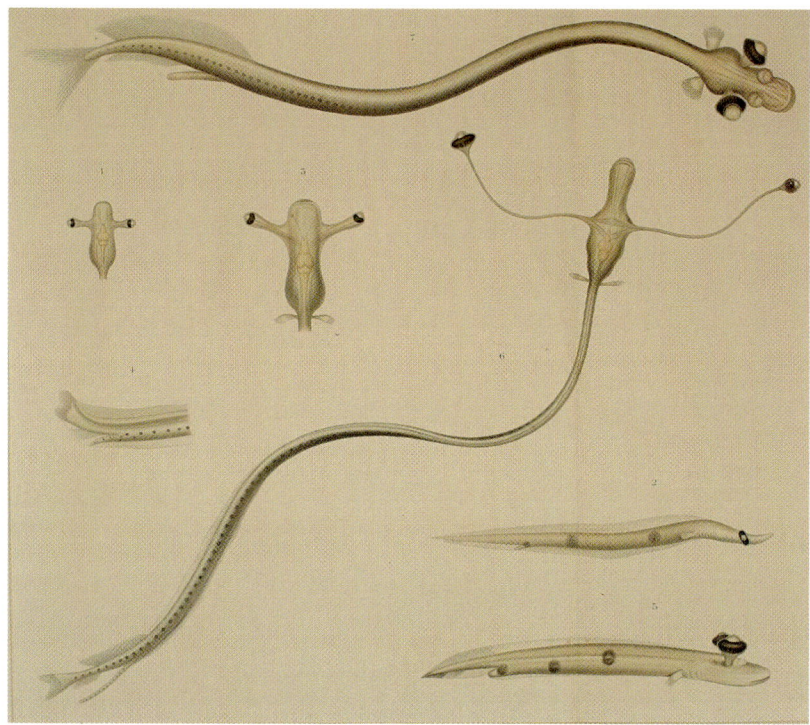

Above: *Larval stage of ribbon sawtail fish, illustration from* Valdivia *expedition report by Carl Chun, vol.* xv *(1906).* Opposite: *Greenland shark, illustration from Jonathan Couch,* A History of the Fishes of the British Islands, *vol.* 1 *(1868 edn).*

avoiding predators. A drawing of an adult ribbon sawtail fish shows that as they mature their eye stalks shrink and get absorbed back into their body, leaving them with normal eyes.

Sharks that lurk in the dark, deep sea have appeared in various artworks and illustrations, some with remarkable accuracy and detail considering that until recently people only knew of them from specimens hauled up from the deep. The Greenland shark (*Somniosus microcephalus*) in Jonathan Couch's *A History of the Fishes of the British Islands* is rather smooth and rounded compared to the craggy-looking animals that can been found almost

3 kilometres (2 mi.) underwater, can grow to more than 5 metres (16½ ft) long and live for hundreds of years. Look into the eye and you'll see a worm-like parasite with a forked tail. This is a type of crustacean (*Ommatokoita elongata*) that commonly latches onto the corneas of these sharks, impairing their vision and likely blinding them, but they survive just fine mostly using their sense of smell to hunt in the darkness for prey, including squid and other sharks.

Also in Couch's book is a bramble shark (*Echinorhinus brucus*), which swims just above the seabed as far as 900 metres (½ mi.) down and is rarely caught by fisheries and even more rarely seen alive. In this drawing, the bramble shark lies on a beach with an anxious round, red eye and a bloated body. Nevertheless, you can see the shark's defining character, the prickly texture of the skin, made from enlarged thorny teeth-like structures called denticles that help protect the shark from attackers. An illustration of a bramble shark in the book *Prodromus of the Zoology of Victoria*, a natural history of the Australian state published in two volumes across the late 1880s by the Irish zoologist Frederick McCoy, shows their dark brown colouration and prickles, which up close look rather like barnacles or little limpets dotted across the body. The less thorny but nonetheless sharply textured skin of angular

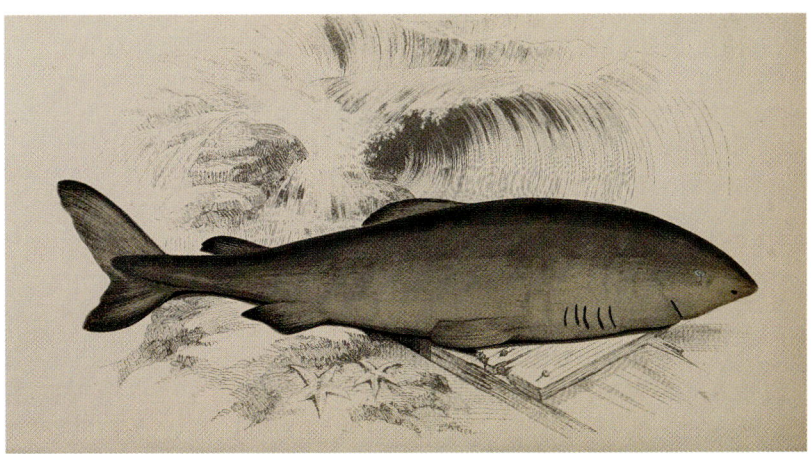

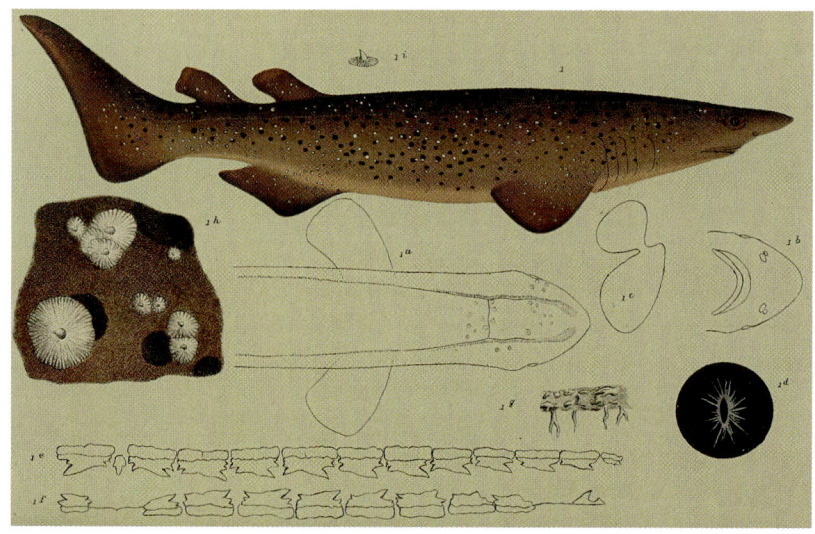

Above: *Bramble shark, illustration by Arthur Bartholomew, from Frederick McCoy*, Prodromus of the Zoology of Victoria, *vol.* II *(1890)*. Below: *Tope shark, illustration from Marcus Elieser Bloch*, Ichthyologie; ou, Histoire naturelle des poissons, *vol.* IV *(1787)*. Opposite: *Ghost shark, illustration from Gottlieb Tobias Wilhelm*, Unterhaltungen aus der Naturgeschichte, *vol.* I *(1799)*.

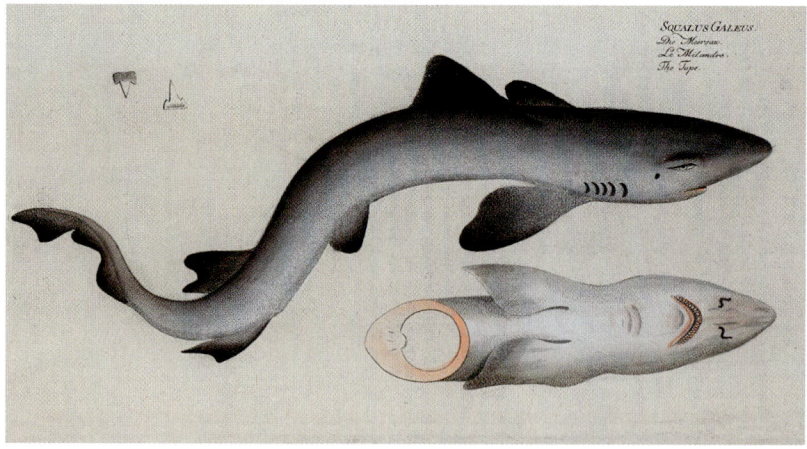

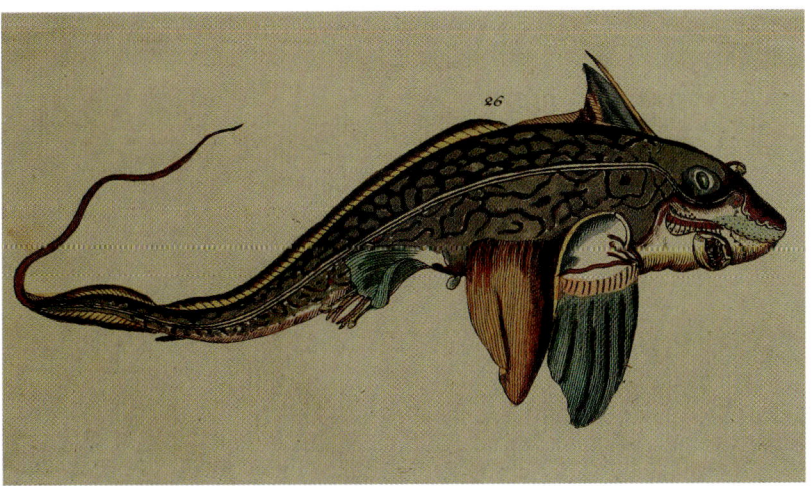

roughsharks (*Oxynotus centrina*) is evident in old drawings, as far back as the mid-seventeenth-century *Historiae naturalis de quadrupedibus libri* (Natural History of Four-Legged Beasts) by the Polish scholar Joannes Jonstonus, and a century later in German naturalist Marcus Elieser Bloch's *Ichthyologie*. Also in Bloch's book is a fabulous drawing of a species that can dive down into the twilight zone, the tope shark (*Galeorhinus galeus*), with a mischievous, side-eyed expression on its face.

Dating from the turn of the nineteenth century, an arresting image appears in the German title *Unterhaltungen aus der Naturgeschichte* (Entertainments of Natural History) by Gottlieb Tobias Wilhelm and engraved by Balthasar Friedrich Leizelt, of an obscure, deep-sea relative of sharks known as chimeras or ghost sharks. These animals have a large head, green eyes and a long body that tapers into a ribbon-like tail. The engraving shows in front of the chimera's dorsal fin a sharp, venom-filled spine, which the fish uses to defend itself from predators. Ghost sharks have wide pectoral fins, which they slowly flap as if they are flying through the deep.

Simple Beauty

A group of far simpler but nonetheless stunning deep-sea animals appears in the *Valdivia*'s immense reports. Sponges don't really look like animals but more like motionless plants or fungi fixed to the seabed. However, they are animals and they sit still and draw water through the pores in their spongy bodies, sifting out particles of food. Animals recognizable as sponges have been living in the ocean for longer than almost all other life forms that live there today. What appear to be fossilized sponges have been found in rocks dating back 890 million years. One volume of the expedition report is devoted to a particular group of these animals known as glass sponges, the hexactinellids, most of which live in the deep sea and are rarely seen in the shallows. Their bodies have sturdy skeletons assembled from tiny pin-like structures, called spicules, with multiple spikes composed of glassy strands of silica. The spicules are differently shaped in each species and many are mesmerizing. Microscope drawings in the *Valdivia*'s report show spicules shaped like six-pointed stars, tiny umbrellas and mushroom caps, and little two-ended boat anchors.

A fold-out page shows intricate illustrations of two whole glass sponges, both with an intricate exterior like lacy mesh. One is candle-shaped and the other is shaped like a French baguette. The *Valdivia* team found the latter off the east coast of Africa, around 1,500 metres (4,900 ft) beneath the waves. The sponge specialist Franz Eilhard Schulze named the species after Carl Chun (*Monorhaphis chuni*). Not featured in the *Valdivia*'s drawing is the immensely long stalk that this sponge would have grown on, made from a single giant spicule that can reach as long as 3 metres (10 ft) – the longest glassy structure made by any known living thing. These glass sponges live exceedingly long lives. In more recent times, scientists have estimated that *Monorhaphis chuni* can live for more than 10,000 years, making them candidates for the longest-lived animals on the planet.

Glass sponges also appear in the artwork of zoologist and artist Ernst Haeckel. One of the hundred prints in his *Kunstformen der Natur* shows a mix of hexactinellid sponges with their lacy exteriors, together with microscope drawings of their spicules. In the lower right corner is the tall, twisted cylinder of a Venus flower basket (*Euplectella aspergillum*), a deep-sea species that partly inspired the design of the City of London skyscraper known as the Gherkin. Opened in 2004, the exterior circular design of the building is not the only aspect borrowed from sponges. The building's internal functions were also inspired by glass sponges and the way they draw water through their bodies to filter food. Instead of water, the Gherkin passively draws in air and circulates it upwards, helping to keep the interior cool and cut down on air-conditioning costs.

The distinctive aesthetic of nature that Haeckel created in *Kunstformen der Natur* was hugely influential on much earlier architecture. From 1890 until the First World War, the Art Nouveau movement was a powerful force in design. Furniture, glasswork, jewellery and buildings were sculpted and decorated in sinuous curves and flowing lines that originated, in part, in Haeckel's studies of plant and animal life. Catalan architect Antoni Gaudí incorporated abstract forms of sea creatures into the stained glass windows, buttresses and arches of the Sagrada Família, the Catholic basilica in Barcelona. The architect of the entranceway to the 1900 Paris World Fair, René Binet, published a book of Art Nouveau interior design, *Esquisses décoratives*, that expanded on his inspirations from Haeckel's studies, in furniture, moulded ceiling roses and ceramic tiling. And many sea creatures flowed through the glass vases of French artist Émile Gallé, who was an important innovator in the French Art Nouveau movement.

The most obvious examples of the organic lines adopted into Art Nouveau come from Haeckel's depictions of various types of jellyfish. These animals were the subject of another milestone publication in Haeckel's scientific career.

Back in 1876, twenty years before the *Valdivia* expedition, a British expedition on board the ship HMS *Challenger* had returned from a circumnavigation of the globe that is widely regarded as founding the discipline of oceanography. During the three-year,

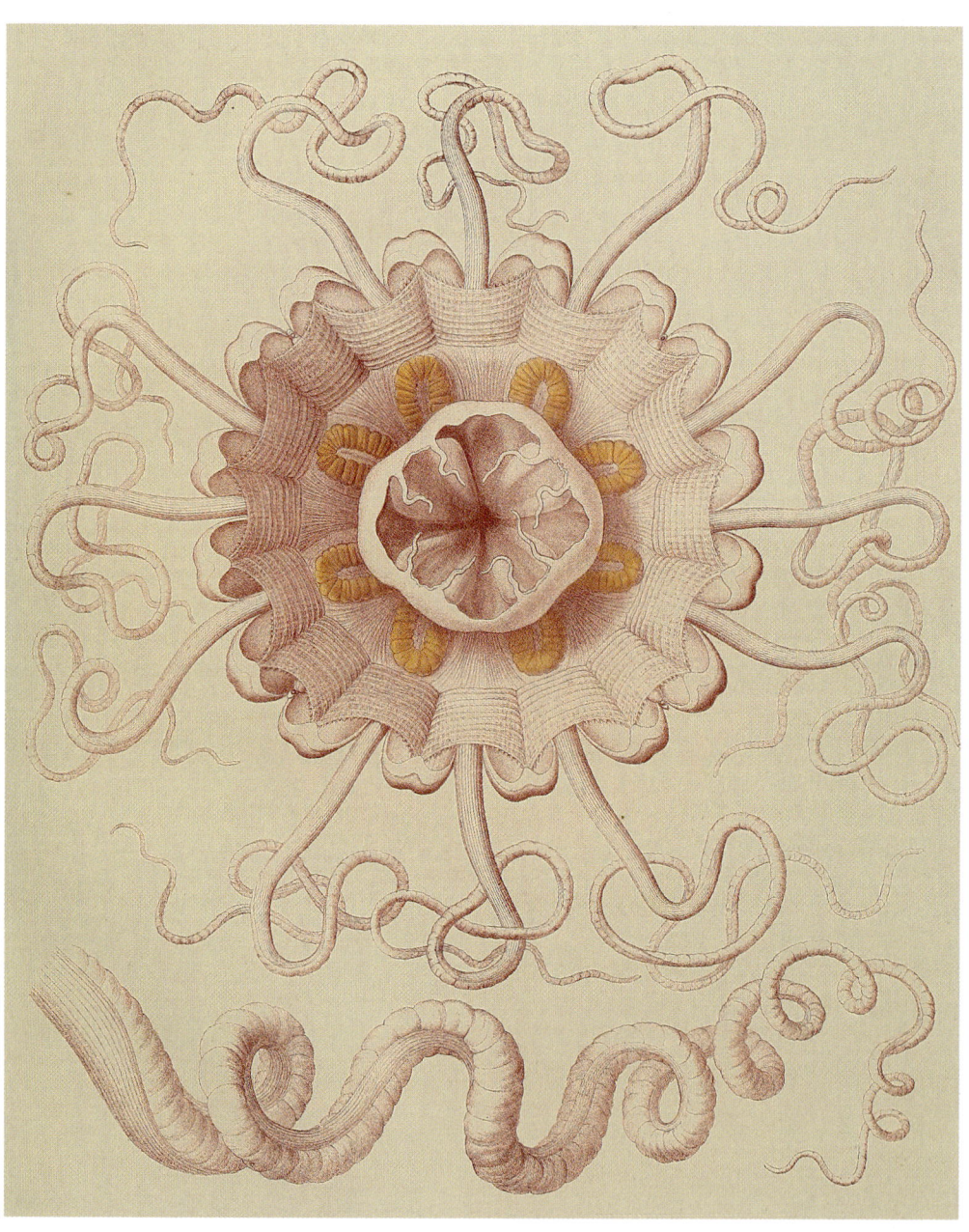

70,000-nautical-mile voyage the team of scientists and sailors measured the depths and contours of the ocean in more detail than ever before, information that was critical for laying the first transoceanic telegraphic cables. The *Challenger* expedition also brought back to Europe thousands of specimens of deep-sea animals, many of them species new to science, which were preserved and distributed among leading experts, who jointly compiled fifty volumes of reports from the expedition. One of those experts was Haeckel. He had not been on board HMS *Challenger*, but afterwards was put in charge of identifying and describing several major portions of the collections, a task that would take him more than a decade. In samples collected from oozy muds, he picked out and described hundreds of microscopic, single-celled radiolarians that had dropped to the deep seabed, their glassy skeletons naturally preserved. The report contains yet more of his detailed drawings, similar to those in his original report on radiolaria.

Haeckel also worked on the *Challenger*'s collections of deep-sea sponges and multitudes of gelatinous creatures that had been carefully scooped up in nets from the open waters of the twilight and midnight zones. These were specimens that were especially poorly understood at the time, and it was up to Haeckel to determine what he was looking at and to give names to any species that he deemed to be new to science – all from preserved specimens that were rarely in pristine condition by the time he saw them.

One of the main challenges for Haeckel as a scientist was to decide where on the tree of life these creatures belonged. In volume XII of the *Challenger* reports, published in 1882, he tackled the expedition's collection of deep-sea medusae, as they were known. Many were true jellyfish (scyphozoans), a taxonomic class named in 1887 by German zoologist Alexander Wilhelm Götte. This group encompasses the jellyfish that live in coastal waters and shallow seas; the likes of moon jellies (*Aurelia aurita*) with

*Helmet jellyfish (*Periphylla mirabilis, *later renamed* Periphylla periphylla*), dorsal view, illustration from the* Challenger *expedition report by Ernst Haeckel, vol. II (1881).*

their smooth, round bodies fringed in fine tentacles; and the giant lion's mane jellies (*Cyanea capillata*) that can trail their tentacles for 30 metres (100 ft) through the water. Haeckel named and described many newly found species of scyphozoans from the deep sea. Some are still recognized today; meanwhile others, such as the helmet jelly (*Periphylla mirabilis*), later taxonomic experts decided were repeats of known species (in this case the jelly *Periphylla periphylla*), a case of an over-zealous species splitting.

Poring over the *Challenger*'s collection of jelly creatures, Haeckel found many life forms that he couldn't easily fit into existing groups of animals. At first sight many look like jellyfish, with their gelatinous bodies, radial symmetry and stinging tentacles dangling down. But he decided that these were different enough that they should be assigned to new categories of life. He named two new taxonomic orders – the narcomedusae and trachymedusae – deep branches of the tree of life that other scientists hadn't previously acknowledged. These types of jelly – such as *Pectyllis arctica* and *Pegantha pantheon* – have smooth-domed bodies ringed in tenacles. Some have long tentacles that look like strings of beads and a tubular mouth with what resembles a bloom of flower petals at the end.

These two groups of jellies, the narcomedusae and trachymedusae, live only in the deep sea and they evolved their own curious ways of surviving in the boundless depths. Their shallow-water cousins, the scyphozoans, produce tiny larvae that sink down and grow into little colonies stuck to the sea floor; periodically these colonies release tiny medusae, the swimming stage of the life cycle, that grow into the animals recognizable as mature jellyfish. The deep-sea jellies don't do this. Instead, because the seabed is usually far out of reach, they carry their larvae inside their bodies until they transform directly into tiny medusae and swim off. These jellies stay pulsing through the water column throughout their whole life cycle, never encountering a hard surface.

*Siphonophore (*Forskalia tholoides*), illustration from the* Challenger *expedition report by Ernst Haeckel, vol.* XXVIII *(1888).*

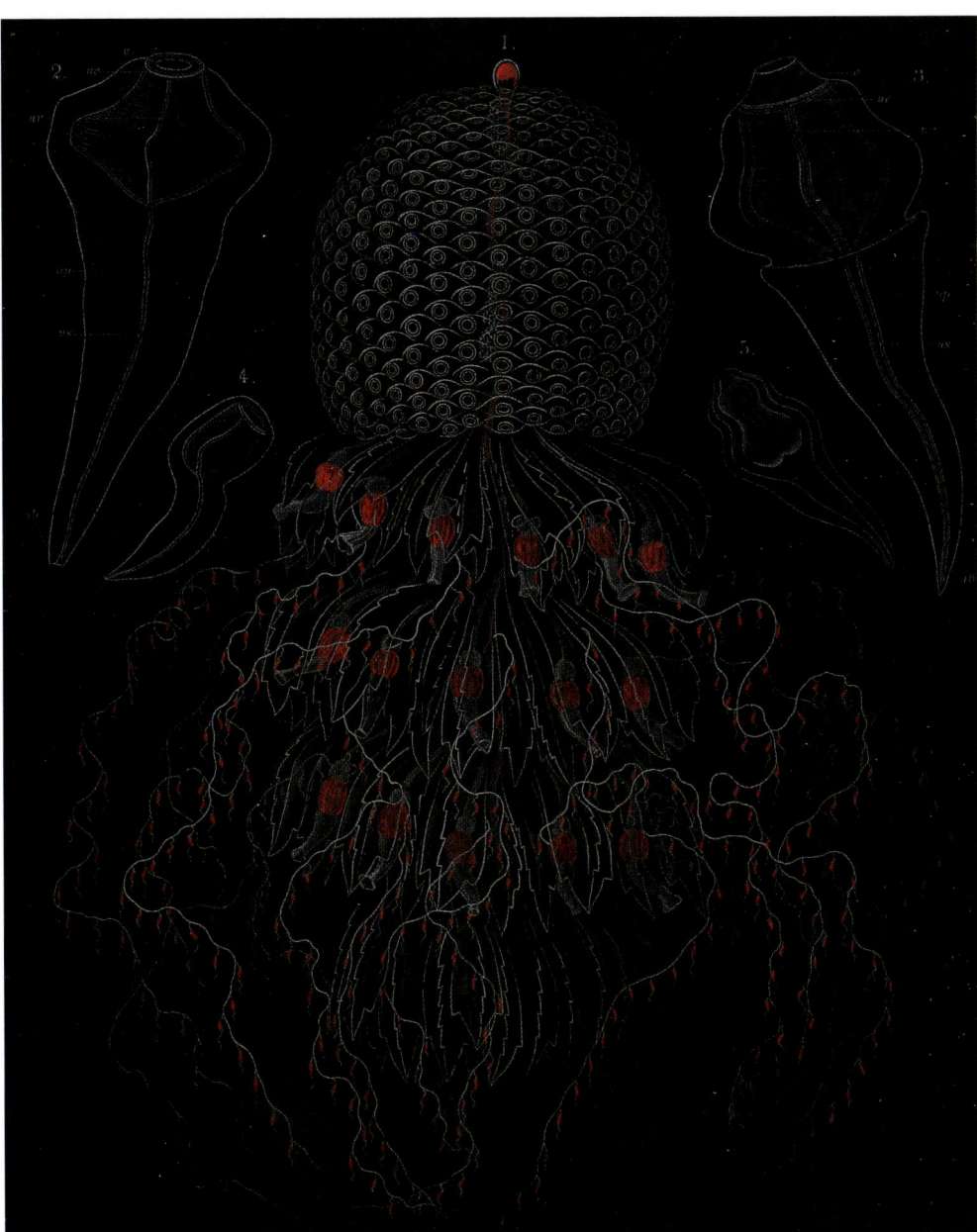

As an artist, Haeckel faced the challenge of producing images of what these animals would have looked like before they were caught and pickled. For this, he inferred much from his own observations of living jellyfish during his studies of coasts and seas around the world. He made illustrations for the *Challenger* report that were forerunners of the beautiful jellyfish plates in *Kunstformen der Natur* that he went on to create in the following decade.

At the library of the Marine Biological Laboratory in Plymouth, I find the 28th volume of the *Challenger* reports. This one, published in 1888, is devoted to another group of delicate, gelatinous animals – the siphonophores, deep-sea relatives of the surface-floating Portuguese man-of-wars. Haeckel himself donated this copy to the library and he signed the frontispiece in elegant script. This volume is illustrated with yet more stunning images of jellyfish-like creatures. Some are white line drawings on black backgrounds, as if Haeckel had drawn them on a blackboard to teach a class of students, although they are far more detailed than would have been possible with a stick of chalk. A personal favourite of mine is *Forskalia tholoides*, a species named by Haeckel that remains scientifically valid today. Their body looks like a pine cone made of neatly arranged eyes, with a ruffled skirt of tentacles.

Haeckel's illustrations have been criticized by some for being too stylized, and that he paid too much attention to aesthetics and arranged animals too neatly. These are, some say, artworks and not scientific studies. However, the images that are now available from deep-diving submersibles of living *Forskalia* show how accurate Haeckel had been in his drawings of these delicate, otherworldly animals, even though he would never have seen one alive.

In all, Haeckel described more than six hundred species of jellies from the *Challenger*'s collections. The multiple volumes of reports that he made added up to the world's first encyclopaedia of jellyfish and their relatives. With his unequalled access to the remote depths, via the *Challenger*'s carefully preserved collections, Haeckel assembled a radical new view of ocean life. He showed that despite the extreme conditions in the deep, diaphanous and beautiful life forms survive down there.

Enter the Aquanauts

Twenty years after Ernst Haeckel's death, a new era of deep-sea study and exploration began when people ventured into this remote realm themselves for the first time. Rather than relying solely on dead or dying specimens hauled to the surface, aquanauts paid brief and dangerous visits to the deep and directly observed its inhabitants in their native habitat.

To survive in the deep, humans depend entirely on technological life support systems, most importantly to keep their bodies protected from the crushing pressure and provide them with breathable air. However, in the first successful version of a deep-sea diving machine these were alarmingly basic. The bathysphere was constructed from a hollow globe of inch-thick cast iron, with three small glass windows, a cramped entrance hatch and an internal diameter of around 1.5 metres (5 ft), just enough space for two trim aquanauts to squeeze into. Oxygen was replenished from two cylinders of compressed gas. Carbon dioxide and moisture were absorbed by a tray of chemicals. To circulate the air inside, the aquanauts brought with them palm fronds to flap. A bright lamp was shone out of one window so they could see what was outside. The metal ball was lowered down into the ocean on a thick metal cable from a winch on the back of a ship. A telephone was installed inside the bathysphere and a cable run up to ship, so the aquanauts could talk to the topside crew. There was no way to control or steer the bathysphere's horizontal position; it could just be lowered down or winched back up. All the aquanauts saw was whatever happened to swim or drift past the windows in the water column.

The two men who invented the bathysphere were the first people to squeeze inside and have themselves lowered into the deep sea. The American zoologist and explorer William Beebe was in his late fifties at the time, accompanied by the younger American engineer Otis Barton. They made their maiden dives inside the bathysphere in 1930, off the island of Bermuda, reaching several hundred metres underwater and entering the ocean's twilight zone.

The pair had already broken record after record for the deepest-diving humans, but it wasn't until 1934 that they achieved their goal of venturing half a mile down (800 metres/2,625 ft). In August that year, they went on a dive to 923 metres (3,028 ft) – they had unspooled the entire cable and could go no deeper. Beebe and Barton had arrived at the fringes of the midnight zone.

By then, Beebe was a celebrity explorer. He published popular books, had broadcast live on national radio from the deep sea inside the bathysphere and wrote about his adventures half a mile down in *National Geographic* magazine. Beebe famously described the moment when he and Barton descended deeper than any conscious humans had ever been (all the others being the dead victims of sea battles and shipwrecks). He wrote, 'We were the first living men to look out at the strange illumination: And it was stranger than any imagination could have conceived.'

These 'strange illuminations' were illustrated in his books and articles by vivid paintings of deep-sea life. These were the work of the German artist Else Bostelmann. The award-winning artist had left Germany after 1909 and emigrated to the United States with her husband, a cello player; there she gave up painting to raise a family. Following her husband's sudden death in 1910, Bostelmann returned to art to support herself and her daughter. She was in her early fifties when Beebe hired her as the expedition artist and she joined the team in Bermuda.

Bostelmann's assignment was to draw and paint the deep-sea creatures that Beebe observed through the cramped windows of the Bathysphere. Photographic cameras at the time were not advanced enough to capture the ephemeral glimpses of animals whizzing past in the darkness. Human eyes accommodated to the dim light

*Sabre-toothed viper fish (*Chauliodus sloanei *now* Chauliodus sloani*) chasing ocean sunfish (*Mola mola*), illustration by Else Bostelmann, 1934.*

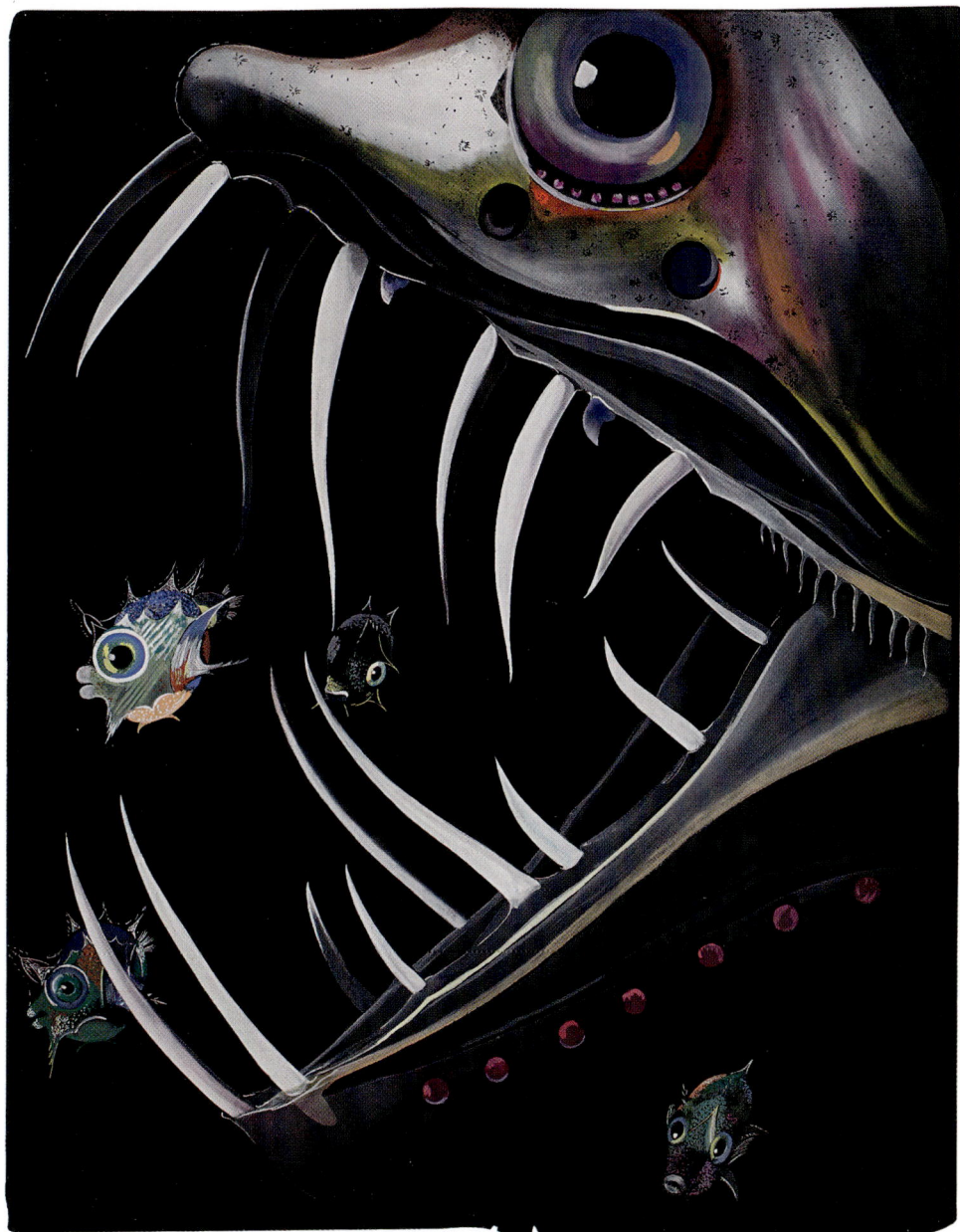

were much better sensors. Bostelmann never dived into the deep herself, apparently because Beebe thought it too dangerous for a mother of a teenage daughter (and while nobody ever came to harm, there were several hair-raising incidents during the bathysphere dives, when the diving chamber flooded, when it crashed into a reef and when the winch cable became twisted). Instead, while he was inside the bathysphere, Beebe picked up the telephone receiver and described what he was seeing out of the window to the team on the ship at the surface. And when the bathysphere was retrieved from the deep and Barton and Beebe were released onto the deck, Beebe would sit down in an artistic huddle, as he called it, with Bostelmann, as she sketched from his notes and memories. Later he wrote, 'Little by little my brain fish materialized.'

One of the most striking and memorable of Bostelmann's paintings shows the open jaws of a viperfish lunging with their sabre-like teeth at a cluster of wide-eyed, star-shaped sunfish larvae. We the viewers are pressed up closed to this hunting scene; perhaps we are one of those young fish in fear of our lives (although, in real life, you'd have trouble poking your finger into the mouth of one of these little fish, *Chauliodus sloani*, which barely grow a handspan in length). This was one of Bostelmann's black paintings, as she referred to them, the dark scenes that imagined, with Beebe's guidance, what these animals look like as they swim through the twilight zone. Another of her black paintings is of a pair of dragonfish that are inspecting the bathysphere and peering into one of the windows at the humans inside.

In another painting, Bostelmann shrinks the viewer into the microscopic world of planktonic swimming sea snails, commonly known as sea butterflies, with their tiny transparent shells and little wings that let them flit underwater. Several other of her images show predation in action – fish chasing strawberry squid, others with shrimp snagged in their jaws – revealing moments from the inner lives of these animals that had been hidden from view for so long.

In all, Bostelmann painted and sketched pictures of hundreds of species from the Bermuda expedition, many based on specimens trawled up in the conventional manner, as her predecessors' had been. The images Bostlemann produced from animals she saw herself share a feeling similar to that of the illustrations of the *Valdivia* expedition; they were carefully observed and accurate to the living species. But when all she had to go on was Beebe's descriptions narrated from the deep and remembered afterwards, the animals have more idiosyncratic and stylized characters – these are ideas of deep-sea animals more than exact depictions of real things, a call back to fairy tales from times when nobody had been into the deep sea.

Bostelmann ventured into the shallower seas of the ocean to see living animals for herself. With Beebe's copper diving helmet resting on her shoulders, she climbed down a ladder, descending to a depth of 10 metres (32 ft), and walked across the seabed. As Zarh Pritchard had done several decades previously, Bostelmann set up a seabed art studio and faced the same challenges of painting underwater. After some trial and error with metal etching plates and pencils, the approach she finally landed on was to use oil paints on a canvas fixed to a music stand held down by heavy weights, her paintbrushes tied with strings so they wouldn't float away. Her scenes faithfully captured the restricted colour palette of the underwater world – all greens and blues, with oranges and reds washed out because the water absorbs those wavelengths from sunlight even in the shallows. She said she preferred to stay within the top 7 metres (23 ft) or so of water and not venture deeper, where the colours, she found, were too few.

Having gained exposure in the public eye from working with Beebe on the Bermuda expeditions, Bostelmann was able to afford her daughter's education and support herself for the rest of her life through her artwork, which continued to focus on nature and the underwater world. She illustrated children's books, created marine-themed textiles and wallpaper, and painted murals for private yachts, sometimes featuring the otherworldly denizens of the deep she had encountered up close in Bermuda.

New Eyes in the Deep

The present day-generation of ocean explorers are spending more time observing the deep than anyone before them. There are good reasons for people to keep going into the deep in person. By bringing back stories of being in the deep ocean they help strengthen connections between this distant realm and human lives, making it exciting and relevant to those who otherwise struggle to care about what's down there. People are also excellent sensors for watching through the dim light of the twilight zone. Engineers have yet to build a camera that is as good as a dark-adapted human eye at catching sight of brief twinkles and flashes of glowing animals whizzing past. The only problem with human eyes is that there is no way (yet) of recording and replaying what they see.

Venturing into the deep in submersible vehicles is an expensive privilege for the few. In a similar way to space exploration, it requires tremendous investment to send people into the deep and funding is coming increasingly from super-rich individuals, with their own agendas, rather than from governments. Only a few vehicles are available to researchers that are capable of taking them to study the ocean's greatest depths, and these fortunate scientists can really only catch glimpses of what exists in this vast and largely unexplored space. Most of what we're learning about the deep ocean and what lives there is coming from robotic devices sent down in place of humans – a far safer and less expensive option.

Remotely operated underwater vehicles, commonly referred to as ROVs, were originally developed by the oil and gas industry to build and fix deep-water drilling installations, then scientists adopted them as a powerful research tools. Usually around the size of a large car, these machines carry bright lights and high-definition cameras; they can have robotic arms fitted and other devices for picking things up, gathering samples and performing simple science experiments. ROVs are lowered into the deep and controlled along miles of cabling that links to a ship at the surface, where people can observe in real time what the robot sees, and instruct it where to go and what to do during dives that last all day.

Some scientists have begun to follow the example of their forebears and are bringing artists on their deep-sea research trips, although the role of these artists has changed. Cameras have replaced the skilled human hands that used to create scientific illustrations of specimens freshly collected from the deep sea, before their colours fade and shapes collapse. Now, the artists are there to produce something else, and to interpret the deep ocean in their own ways.

The Schmidt Ocean Institute's artists-at-sea programme opens a dialogue between scientists and artists who jointly explore remote reaches of the deep ocean on board the research vessel *Falkor (too)*. The naming of the vessel, and its predecessor, after Falkor the luck dragon in the 1979 novel and various screen adaptations of *The Neverending Story* by Michael Ende, reflects the never-ending journey of the institute on its mission to explore the ocean's vast ecosystem. The artists and scientists on *Falkor (too)* have a shared view of the deep thanks to ROV *SuBastian*, a remotely operated vehicle named after the human protagonist in *The Neverending Story*, which can dive to 4,500 metres (nearly 3 mi.) underwater. The artists work alongside scientists and create paintings, drawings, prints, weavings and sculptures while out at sea, beyond the horizon and floating above the abyss. Some artists focus on the physical nature of the deep, the geography and geology of hidden seascapes that are being mapped for the first time; they interpret the scientific data and incorporate it into their work, the depths, temperatures and currents measured by electronic devices. In 2018 Lori Hepner created digital artworks from traces of LED lights fixed to her body while she moved and danced on the deck of the ship, interpreting the seascape below. Other artists concentrate on the life revealed in the deep during these explorations, such as Carlos Hiller's 2023 paintings of a single white deep-sea skate gleaming in the darkness as it swims past a giant underwater escarpment, and another of a newly discovered deep-sea octopus nursery off the coast of Costa Rica, where hundreds of females congregate in the warm waters seeping from a deep undersea mountain to incubate their unhatched eggs.

Images captured by robots are themselves compelling works of art, impressions of the secret world of the deep briefly interrupted by the machine passing through. A photograph of a whale skeleton lying on the seabed off the California coast, in the Monterey Bay National Marine Sanctuary, has the quality of a Dutch master still-life, in its dark background and the luminous detail of the subjects. Scientists on board the Ocean Exploration Trust's ship E/V *Nautilus* found the remains of this whale in 2019. This is what's known as a whale fall, an ecosystem that assembles around the dead body of a whale that sinks into the deep and brings an important feast of food to the hungry deep, one that can sustain hundreds of species for years and even decades. Video of what the scientists were seeing was being broadcast in real time over the Internet and people all around the world were watching as the ROV's cameras panned across the scene – itself an incidental work of performance art. The photograph is a still shot from that video feed. Dozens of scavenging animals, including crabs, eelpout and mauve-coloured octopuses are picking the skeleton clean. A red, fuzzy covering on the skeleton is made up of thousands of bone-eating worms, *Osedax* (from the Latin words *os* and *edax* meaning 'bone' and 'devourer'). Scientists first identified the puzzling life form from a whale fall, also in Monterey Bay, back in 2002. These animals look more like plants than worms; they have no mouth and no guts; they have a green root-like structure that burrows into bones helped by acid secretions, and a flowery tuft of red gills sticking up to breathe oxygen from the water.

Osedax worms have only ever been seen on the bare bones of vertebrates lying on the deep seabed, and not just whales but also tuna, turtles, sharks and even dead alligators that scientists deliberately sank into the abyss to see what would come and find them. Dozens of species have been found on carcasses dotted across the ocean, from the tropics to the icy polar seas. All of these worms have come to be bone specialists. They eat nothing else. How the worms find bones to settle on remains a mystery, the blank part of their lives in between the busy oases of food in the abyss.

American artist Ellen Gallagher and Dutch photographer Edgar Cleijne were inspired by the original discovery of *Osedax* worms to create an animated film of the same name, part of a 2013 installation at the New Museum in New York. The film takes viewers on a journey that explores the world with no plan in mind of what will be found and at first no clue what they're looking at, a mindset shared by all deep-sea scientists who learn that they must always keep their minds open to unexpected encounters. Finding the sunken body of a dead whale, covered in previously unimagined life forms, was not on anyone's agenda that day when *Osedax* first came to light.

Gallagher and Cleijne's film features a shipwreck, human faces made partly of sea life and a whale falling to the seabed, taking with it all the knowledge that it accumulated as it swam through the ocean during its long life. Then the small *Osedax* worms inscribe their own layers of knowledge and understanding into the bones left behind by these giants. Matter in constant motion and transformation is a central theme in the work. Gallagher recognizes an image of diaspora in the worms that

Submerged whale skeleton covered in octopuses, Monterey Bay National Marine Sanctuary.

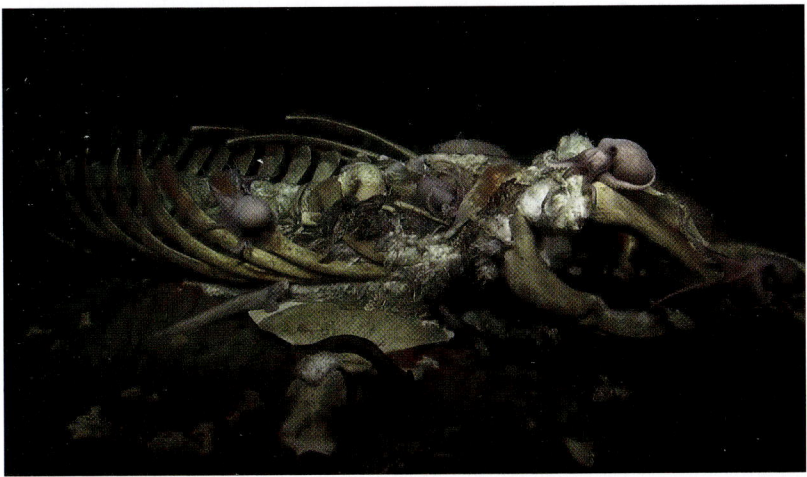

travel unseen across the ocean, each whale fall forming a satellite for another. Her view is deeply ecological, of matter and energy passing between forms and forging distant connections.

Even with so many scientists and artists now focusing their gaze on the deep, still we are learning more about it all the time. Humans will never know everything there is to know about life in the great, vast ocean, a reality we should celebrate and defend in any way we can. There are those today who are thirsty to exploit the deep sea for its mineral resources and edible creatures. Their view for the future of the deep is to convert this, the planet's most untouched and mysterious wilderness, into a lifeless industrial zone, and all in the name of corporate profit and personal greed. Thankfully, more people than ever are seeing things differently. Instead of tearing it apart, we can keep on learning about the entire ocean, worm by worm, fish by fish, safe in the knowledge that there will always be more discoveries to make and new ways of looking at and thinking about this extraordinary place.

FURTHER READING

Atkins, Anna, *Cyanotypes* (Cologne, 2023)

Cohen, Margaret, 'Seeing through Water: The Paintings of Zarh Pritchard', in *Coastal Works: Cultures of the Atlantic Edge*, ed. Nicholas Allen, Nick Groom and Jos Smith (Oxford, 2017), pp. 205–24

DeCaires Taylor, Jason, *The Underwater Museum: The Submerged Sculptures of Jason deCaires Taylor* (San Francisco, CA, 2014

Duffin, Christopher J., 'Natternzungen-credenz: Tableware for the Renaissance Nobility', *Jewellery History Today*, XIV (2012), pp. 3–5

Forli, Maurizio, and Andrea Guerrini, 'Scorpions, Ants, and Other Stone Insects: The Understanding of Trilobites over the Centuries', in *The History of Fossils over Centuries*, ed. Maurizio Forli and Andrea Guerrini (New York, 2022), pp. 417–42

Fox, Brad, *The Bathysphere Book: Effects of the Luminous Ocean Depths* (London, 2023)

—, 'Drawing the Deep Sea from a Seat on the Shore', *Hakai*, www.hakaimagazine.com, 16 May 2023

Haeckel, Ernst, *Art Forms from the Ocean: The Radiolarian Prints of Ernst Haeckel* (Munich, 2005)

Harvell, Drew, *A Sea of Glass: Searching for the Blaschkas' Fragile Legacy in an Ocean at Risk* (Oakland, CA, 2016)

Irmscher, Christoph, and Richard J. King, *Audubon at Sea: The Coastal and Transatlantic Adventures of John James Audubon* (Chicago, IL, 2022)

Jones, Erika, *The Challenger Expedition: Exploring the Ocean's Depths* (London, 2022)

Jovanovic-Kruspel, Stefanie, Valérie Pisani and Andreas Hantschk, '"Under Water" – Between Science and Art – The Rediscovery of the First Authentic Underwater Sketches by Eugen von Ransonnet-Villez (1838–1926)', *Annalen des Naturhistorischen Museums in Wien. Serie A für Mineralogie und Petrographie, Geologie und*

Paläontologie, Anthropologie und Prähistorie, CXIX (2017), pp. 131–53

Kemp, Christopher, *Floating Gold: A Natural (and Unnatural) History of Ambergris* (Chicago, IL, 2012)

McNamara, Kenneth J., *The Star-Crossed Stone: The Secret Life, Myths, and History of a Fascinating Fossil* (Chicago, IL, 2010)

Mason, Adrienne, 'An Ocean in the Parlor', *Hakai*, www.hakaimagazine.com, 24 March 2017

Miyazaki, Yusuke, and Murase Atsunobu, 'Fish Rubbings, "Gyotaku", as a Source of Historical Biodiversity Data', *ZooKeys*, CMIV (2019), pp. 89–101

Mora, Camilo, et al., 'How Many Species Are There on Earth and in the Ocean?', PLOS *Biology*, IX/8 (2011), e1001127

Morse, Kate, 'Shell Beads from Mandu Mandu Creek Rock-Shelter, Cape Range Peninsula, Western Australia, Dated before 30,000 BP', *Antiquity*, LXVII (1993), pp. 877–83

Moyle, Peter B., and Marylin A. Moyle, 'Introduction to Fish Imagery in Art', *Environmental Biology of Fishes*, XXXI (1991), pp. 5–23

Probyn, Elspeth, 'How to Represent a Fish?', *Cultural Studies Review*, XXIII/I (2017), pp. 37–59

Staaf, Danna, *The Lady and the Octopus: How Jeanne Villepreux-Power Invented Aquariums and Revolutionized Marine Biology* (Minneapolis, MN, 2022)

Vanselow, Klaus Heinrich, Sven Jacobsen, Chris Hall and Stefan Garthe, 'Solar Storms May Trigger Sperm Whale Strandings: Explanation Approaches for Multiple Strandings in the North Sea in 2016', *International Journal of Astrobiology*, XVII/4 (2018), pp. 336–44

Voss, Julia, and Rainer Willmann, *The Art and Science of Ernst Haeckel* (Cologne, 2017)

Widder, Edith, 'The Fine Art of Exploration', *Oceanography*, XXIX/4 (2016), pp. 170–77

Williams, Tracey, *Adrift: The Curious Tale of the Lego Lost at Sea* (Lewes, 2022)

ACKNOWLEDGEMENTS

This copy of *Ocean Art* is in your hands, on your screen or perhaps in your ears thanks entirely to the support and patience of Michael Leaman, who has so kindly given me the freedom to explore these interwoven topics. My thanks to all the team at Reaktion Books, especially Alex Ciobanu, Amy Salter, Helen McCusker and Fran Roberts.

The making of this book about marine biology and art owes a great deal to the people who over the years have encouraged, instructed and assisted me in either discipline. My thanks to everyone who has helped me to appreciate and understand the joys and wonders of the living world, in particular Joy Percy, Al Edwards, Laurie Friday, Adrian Friday, Nick Davies and Peter Herring. For my own explorations in visual art, thank you to James Hill and Fiona Thomas at St Barnabas Press for helping me to make an enjoyable mess in the printmaking studio. And my heartfelt thanks to the artists who never fail to amaze and inspire me in their creativity, especially Aaron John Gregory, Rômolo D'hipólito, Good Wives and Warriors, Nicola Davies and Jackie Morris.

As an author, I am endlessly grateful to Simon Winchester and Emma Sweeney for their pivotal help in those early, difficult days of my wanting to write books. Even earlier than that, my mother tells me there was a primary school teacher, when I was around six or seven years old, who saw something of the nature writer in me (as well as an untameable rambler). So, here's to the memory of dear Mrs Elmer, who is surely long gone by now.

In more recent times, I am thankful on a daily basis for all the science and nature writers with whom I share remote work-spaces and exchange virtual office chat, including Ben Hoare, Isabel Thomas, Jess French, Jules Howard, Liam Drew, Emma Bryce, Anna Politano, Helena Pozniak, Sarah Wild and Manuela Callari.

And, as always, my thanks and love to all my family and friends who continue to celebrate me, my words and books, and especially to Ivan for being my companion in art museums, on waves and so much more.

PHOTO ACKNOWLEDGEMENTS

The author and publishers wish to express their thanks to the sources listed below for illustrative material and/or permission to reproduce it. Some locations of artworks are also given below, in the interest of brevity:

Alamy Stock Photo: pp. 30 (Staatliche Kunstsammlungen Dresden/INTERFOTO), 106 (Randy Duchaine); © ARS, NY and DACS, London 2024, photo Bernard O'Kane/Alamy Stock Photo: p. 182; Art Institute of Chicago: pp. 107 *below*, 174; Boston Public Library: pp. 162, 172, 200, 201; Brooklyn Museum, New York: p. 154; © Center for Creative Photography, The University of Arizona Foundation/DACS, London 2024: p. 53; The Cleveland Museum of Art, OH: pp. 9, 20, 78, 80, 81 *below*, 125; © DACS, London 2024, courtesy Jason deCaires Taylor: p. 156; © Salvador Dalí, Fundació Gala-Salvador Dalí/DACS, London 2024, photo Tate, London: p. 113; Ernst Mayr Library of the Museum of Comparative Zoology, Harvard University, Cambridge, MA: pp. 94, 95, 96, 97 *above* and *below*, 217, 220; ETH-Bibliothek Zurich: pp. 69, 187; The Fitzwilliam Museum, University of Cambridge: p. 67; Flickr: p. 126 (Museo Archeologico Nazionale, Naples; photo Carole Raddato, CC BY-SA 2.0); Funabashi City Nishi Library: p. 202; courtesy Galerie Zacke, Vienna/www.zacke.at (auctioned 22 April 2022, lot 209): pp. 121 *below*, 122; George Peabody Library, Johns Hopkins University, Baltimore, MD: p. 84; Getty Research Institute, Los Angeles: pp. 85, 87; Gottfried Wilhelm Leibniz Bibliothek, Hannover: p. 90; Indianapolis Museum of Art at Newfields, IN: p. 148; John M. Kelly Library, University of Toronto: p. 195; Kunsthistorisches Museum, Vienna: p. 128; Nick Kwan/Unsplash: p. 43; Library of Congress, Prints and Photographs Division, Washington, DC: pp. 14, 19, 36, 188, 189, 198; Los Angeles County Museum of Art (LACMA): p. 12; Mary Evans/Natural History Museum, London: p. 185; MBLWHOI Library, Woods Hole, MA: pp. 92, 93, 135, 138, 168, 210, 211, 213, 214,